Stereochemistry

An Introductory Programmed Text

Roger W. Giese
Northeastern University
Boston, Massachusetts

Robert P. Mikulak
U.S. Disarmament Commission
Washington D.C.

Olaf A. Runquist
Hamline, University
St. Paul, Minnesota

Burgess Publishing Company
Minneapolis, Minnesota

Printed in the United States of America
ISBN 0-8087-0745-0
Library of Congress Card Number 75-29853

0 9 8 7 6 5 4 3 2 1

Reproduced from Author supplied copy.

Contents

Page

Section 1: Introduction

Many organic molecules differ only in the spatial arrangement of their atoms. Such molecules can share both certain chemical and physical properties, and yet be widely divergent in others. The study of the ways in which the chemical and physical properties of molecules are influenced by the spatial arrangements of their atoms is the study of stereochemistry.

The objective of the first section (the following six questions and answers) is to give the reader some idea of the general relevance and importance of this subject, particularly in regard to life processes.

Introduction

Q-1 The two sea shells pictured here are alike in all respects except one. What feature distinguishes them from one another?

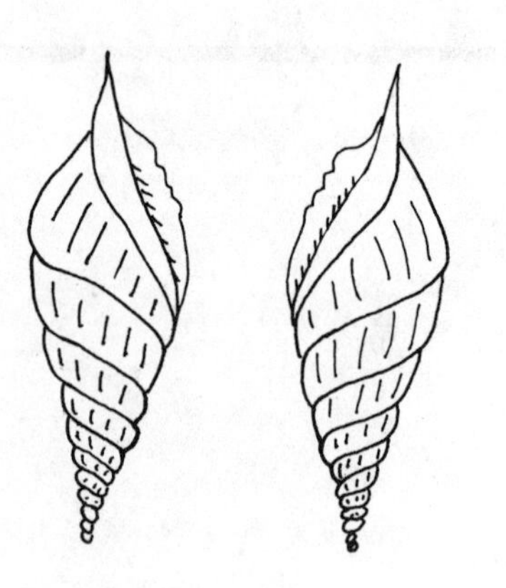

A-1 The shell on the right is a right-handed spiral. The one on the left is a left-handed spiral. This is a good example of macroscopic stereochemistry. The two shells are mirror images of one another, but are not superimposable, just like your hands.

Although the lightning whelk of the Carolinas and Texas is normally a left-handed spiral, most coiled shells have a right-handed spiral. However, sometimes a typically right-handed species will produce a left-handed shell.

The macroscopic stereochemistry of sea shells has not yet been explained on a chemical basis. It is likely, however, that the visible stereochemical features are determined by stereochemical properties at the molecular level.

Q-2 The two compounds shown below can provide an excellent example of the important effect of molecular shape on chemical and physical properties.

(1) cis: H and CO_2H on upper C; H and CO_2H on lower C, same side: $HC(CO_2H)=C(H)CO_2H$

(2) trans: H and CO_2H on upper C; HO_2C and H on lower C.

In what way do the structures of these two compounds differ?

A-2 In (1) the CO_2H groups are on the same side of the molecule, in (2) on opposite sides. The same is necessarily true for the hydrogen atoms.

Q-3 In spite of the similarity in the structures of (1) and (2), they have very different physical properties:

	(1)	(2)
m. p.	403 K	560 K
solubility in 100 cm^3 H_2O, 298 K	78.8 g	0.7 g

How can these differences be explained?

A-3 In order for a crystal to melt or dissolve, the crystal lattice must be physically disrupted. The stronger the self-association of the molecules forming the crystal lattice, the more difficult it will be to disrupt the lattice structure. In other words, the stronger the self-association, the higher the melting point and the lower the solubility. Apparently the shape of (1) leads to much stronger self-association in the crystal lattice than does the shape of (2).

Q-4 Bombykol, the sex attractant of the silk moth, has the following general structure:

$$\overbrace{CH_3(CH_2)_2}^{R}CH=CH-CH=CH\overbrace{(CH_2)_8CH_2OH}^{R'}$$

The biological activities of the four compounds with this general formula are given below:

Compound	Relative Activity
(1) H(R)C=C(H)–C(H)=C(H)R' (R and R' cis to their neighbouring chain; H/H, R/H)	1
(2) R(H)C=C(H)–C(H)=C(R')H	10^3
(3) H(R)C=C(H)–C(H)=C(H)R' (R' trans)	10^{12}
(4) R(H)C=C(H)–C(H)=C(H)R'	10^{-1}

What feature of bombykol obviously has a major effect on its sex attractant power?

A-4 Clearly the activity is in part dependent on the precise orientation of the atoms in space. This is one of the examples of the importance of stereochemistry in biological processes.

Q-5 Compounds (5) and (6) are quite similar in appearance except for two features. What are these features?

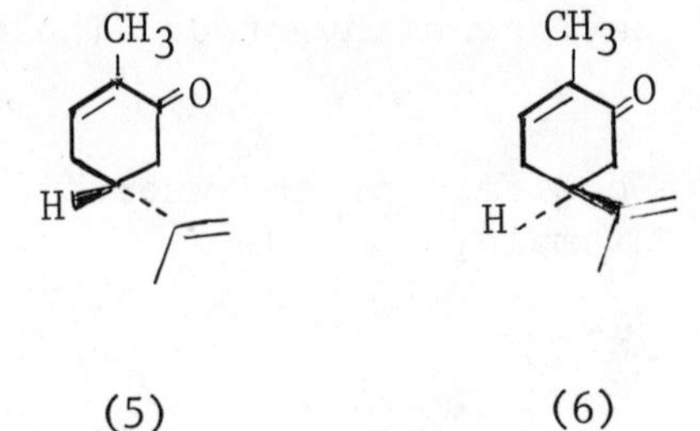

A-5 1. The H atom is joined to the six-membered ring by a wedge in (5) and a set of dashed lines in (6).

2. Just the opposite is true for the " ⟩= " group. The wedge and dashed lines are stereochemical conventions which will be fully explained later. They indicate that the two groups have exchanged their spatial position in the two compounds

Q-6 The two compounds (5) and (6) have the same boiling point and solubilities because the spatial arrangements of their atoms are so similar. Nonetheless, they have distinctly different odors. Compound (5) has a strong, sharp spearmint odor, while compound (6) has the odor associated with caraway seeds. What does this tell you about the odor sensing system of the body?

A-6 Apparently the system which the body uses to distinguish odors is exquisitely sensitive to molecular shapes.

Another example of this phenomenon is provided by compounds (7) and (8).

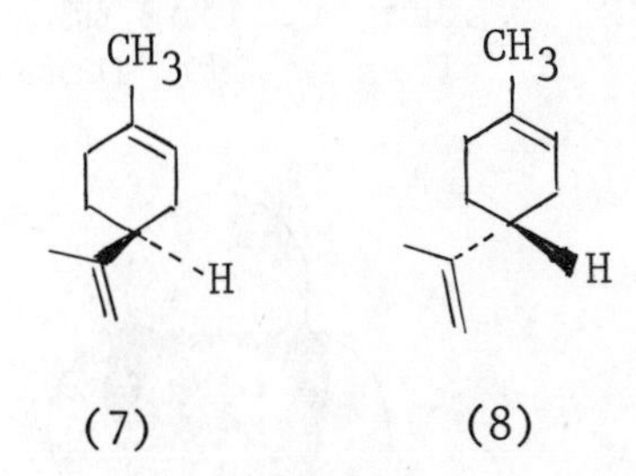

Compound (7) has the odor of lemons while (8) is responsible for the odor of oranges.

Models and Perspective Drawings:

S-1 Inspection of a molecular structure often leads to valuable insights into the physical and chemical properties of the molecule. Since an individual molecule is extremely small and therefore difficult to observe directly, several types of mechanical models have been invented. The models are of two general types:

1. skeletal models which show only the position of the nuclei and the bonds connecting them, for example,

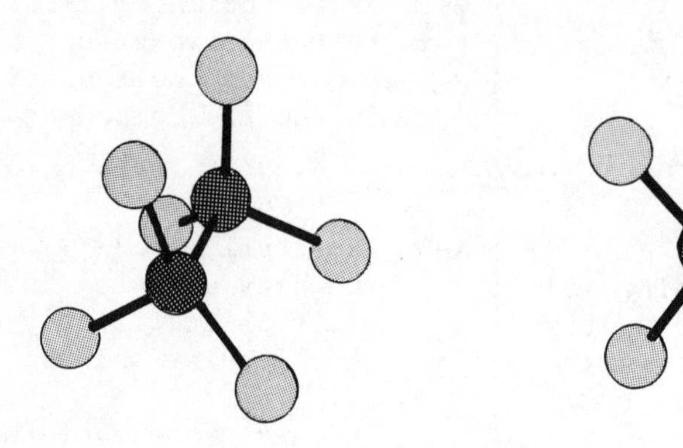

2. space-filling models which indicate the actual relative sizes of the atoms, for example,

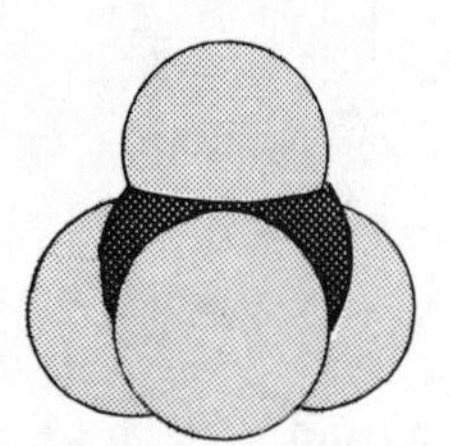

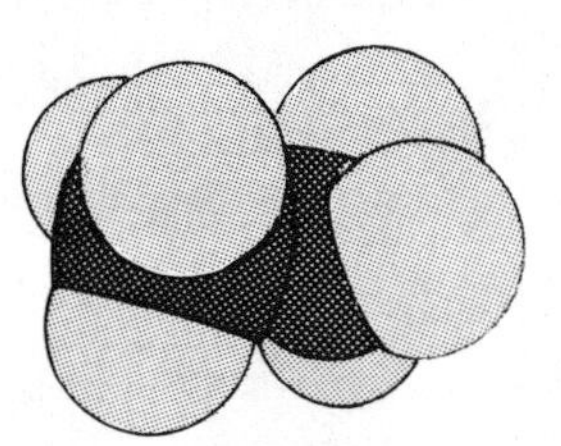

In spite of the great usefulness of models in helping to visualize molecular structure, you must remember that models greatly oversimplify the actual structure. For example, models often do not respond to strain the way an actual molecule would. In general, models are too stiff in resisting angle bending, too loose in rotating around single bonds, and too intolerant of atomic compression (particularly when the models are the space-filling type).

Q-7 Which type of model would be most useful for checking if two groups will "get in each other's way?" (Atoms cannot occupy the same place at the same time.)

A-7 A space-filling model would be most useful since it shows the relative sizes of the atoms.

Q-8 Which type of model would best show bond angles and bond lengths?

A-8 A skeletal model, which shows only the positions of the nuclei and the bonds between them.

S-2 A three-dimensional model gives more information on spatial relationships than a perspective drawing can. However, a drawing is often better suited for particular purposes, such as when a model would be awkward to use, cannot be used at all, or simply is not necessary. Chemists have adopted a number of conventions for translating a three-dimensional structure into a perspective drawing.

One commonly used convention is the "flying wedge." Bonds to groups which lie in the plane of the paper are shown as solid lines. Wedges denote bonds to groups situated above the plane of the paper. Dashed lines denote bonds located below the plane. The drawing below illustrates one of a number of possible "flying wedge" representations for 2-bromopropane.

"flying wedge" drawing

(1)

Q-9 In the "flying wedge" drawing shown below, which groups are in the plane of the paper? Above? Below?

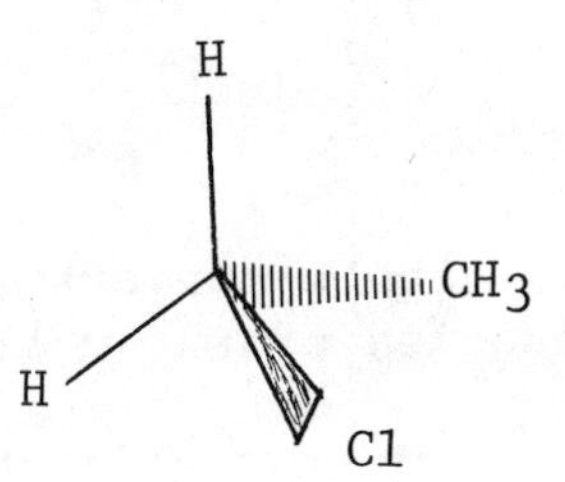

A-9 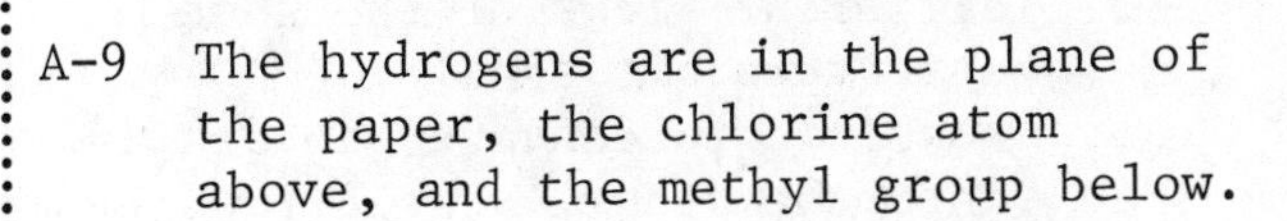
The hydrogens are in the plane of the paper, the chlorine atom above, and the methyl group below.

Q-10 Draw a "flying wedge" projection of tetrachloromethane with the atoms oriented as in (1).

A-10

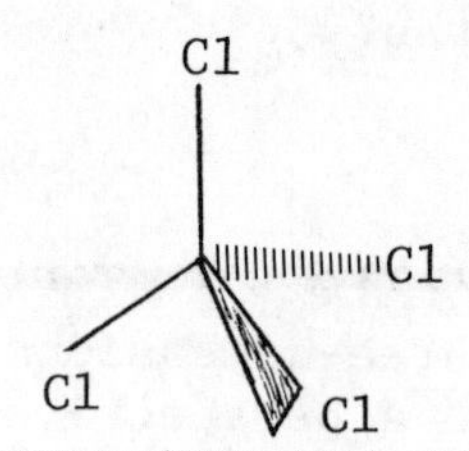

Q-11 Draw a "flying wedge" projection of trichloromethane by substituting a hydrogen atom for a chlorine atom in the drawing shown in A-10. Place the hydrogen above the plane of the paper.

A-11

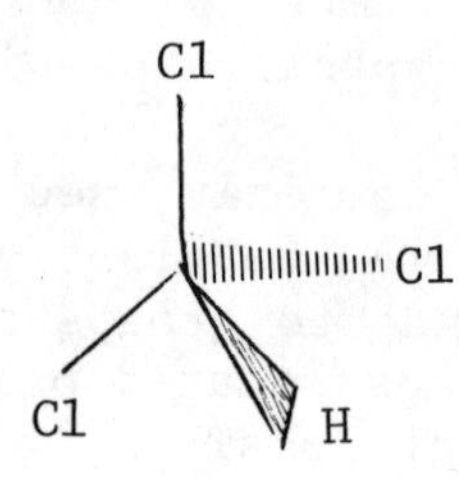

Q-12 Draw a "flying wedge" projection of trichloromethane again, but this time place the hydrogen atom in the plane of the paper.

A-12

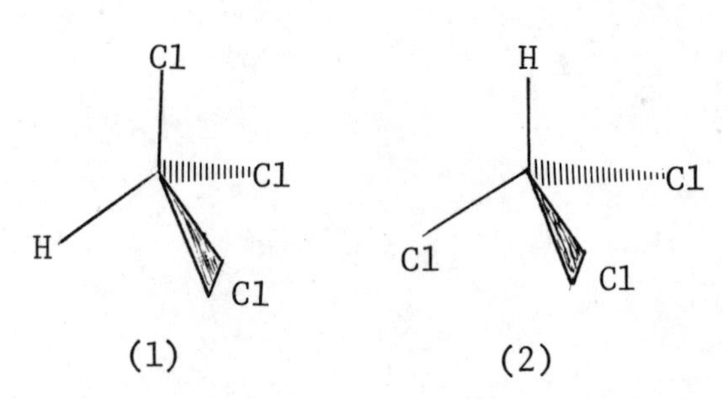

(1) and (2) are both correct. They are different views of the same structure.

Q-13 Since (1) and (2) are merely different views of the same structure, it should be possible to convert one drawing into the other. Use arrows to show how (1) can be converted to (2). Check your answer using models.

A-13

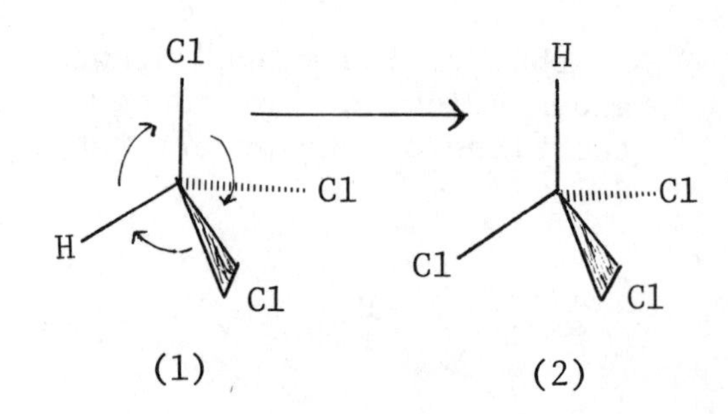

(1) can be converted to (2) by rotation of the molecule.

Q-14 What is the relationship among (1), (2), (3), and (4)? Can they be interconverted?

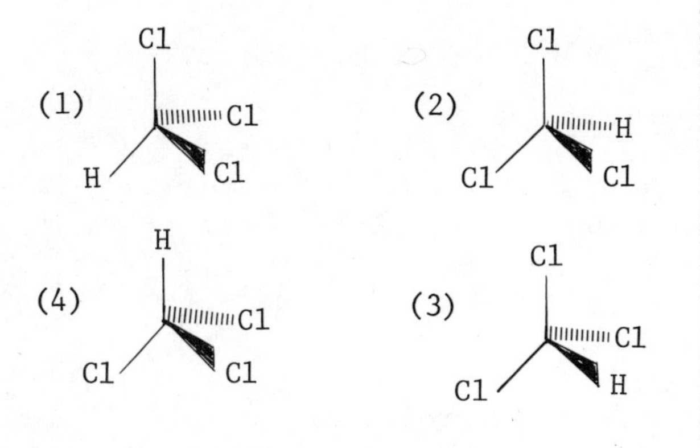

A-14 (1), (2), (3), and (4) are different views of a $CHCl_3$ molecule. They can be interconverted by rotation around one bond. Use models to check this.

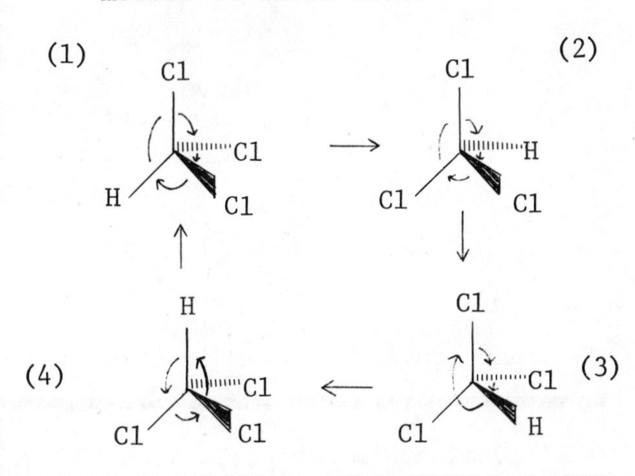

Q-15 Draw the three different views of the molecule below which are obtained by rotation around the bond from D to the central carbon.

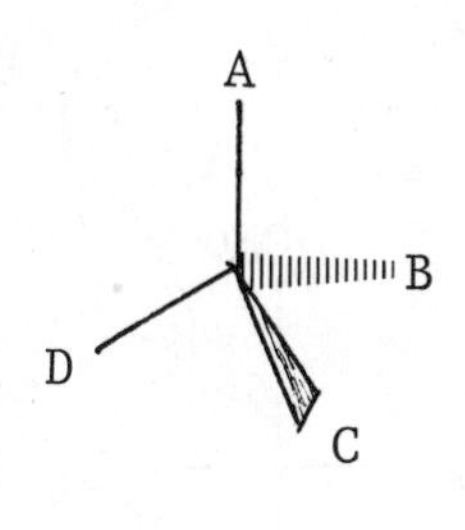

A-15

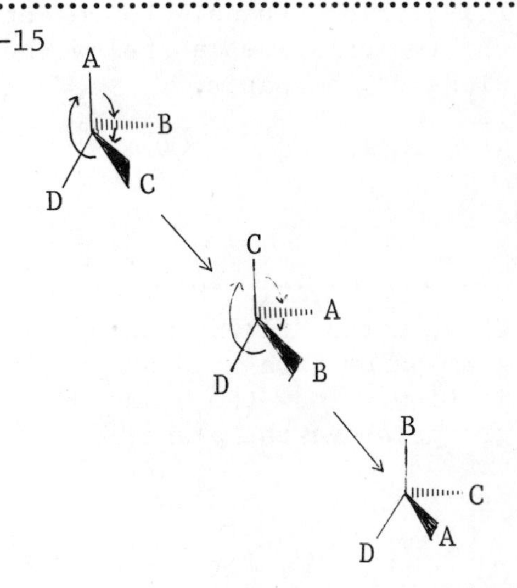

Q-16 Which of the drawings below uses the "flying wedge" convention?

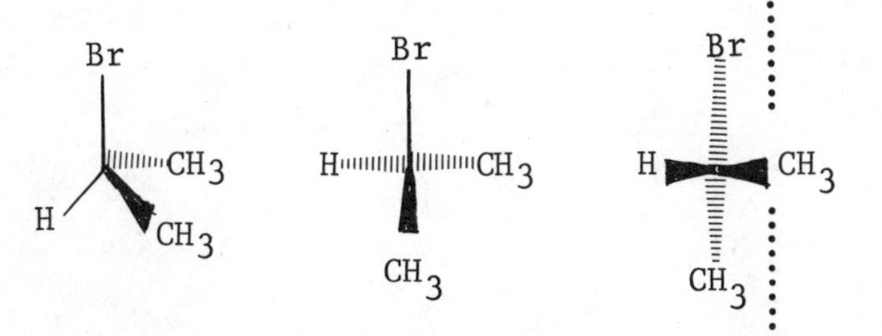

A-16 They all do. Although the orientations of the groups differ, the same convention for indicating position with respect to the plane of the paper is used for each drawing.

S-3 Another useful convention employs a "cross" notation in which horizontal lines represent bonds above the plane of the paper, vertical lines represent bonds below the plane.

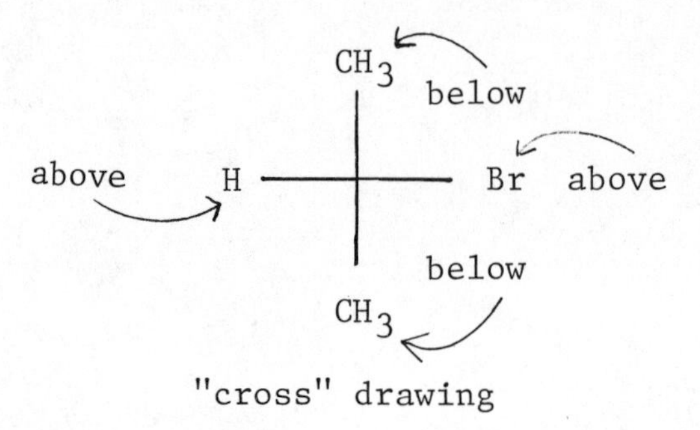

"cross" drawing

This particular notation is often called a "Fischer projection."

Q-17 Draw a Fischer projection of 1,1-dichloroethane. Represent the two chlorines as below the plane of the paper.

A-17

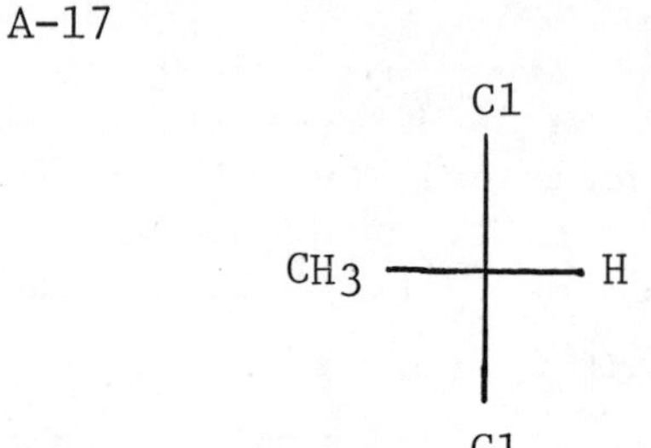

Q-18 Convert the "flying wedge" diagram below to a "cross" projection. Show the H and OH groups above the plane.

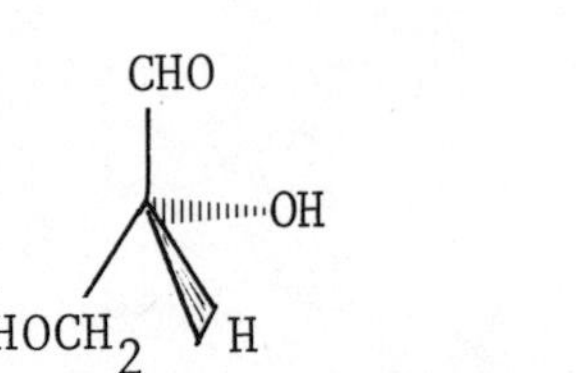

A-18

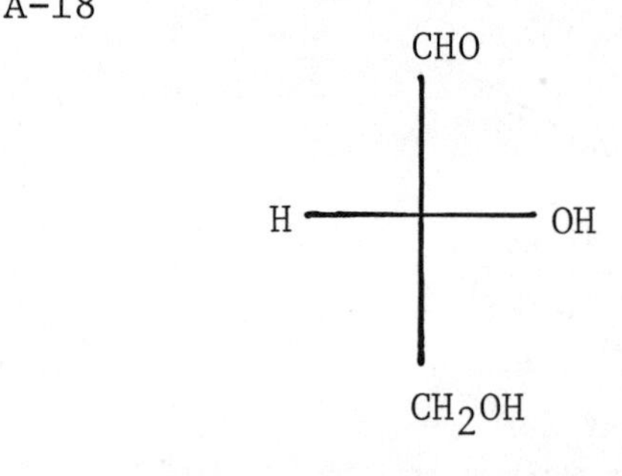

Q-19 Manipulations of the Fischer projection are permitted as long as this drawing is not lifted out of the plane of the paper during the process. Is this manipulation a valid one?

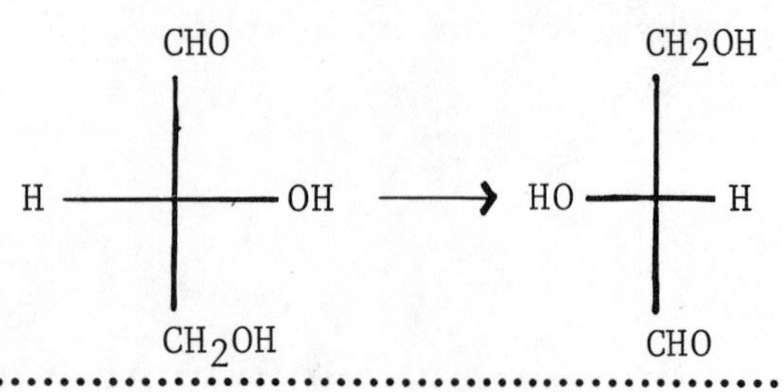

A-19 Yes, since the drawing was rotated within the plane of the paper.

Q-20 Is the manipulation shown below a valid one?

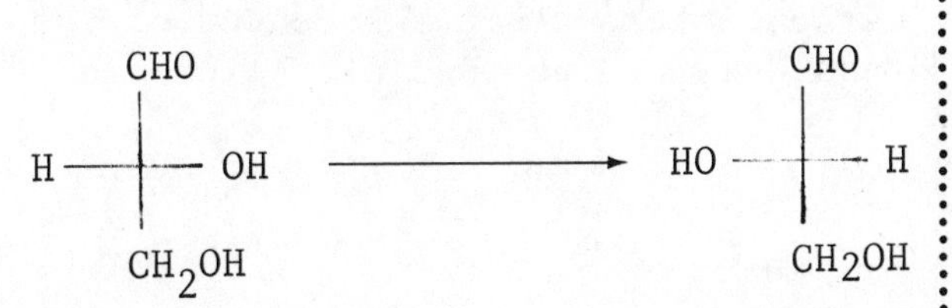

A-20 No, it is not. The drawing was lifted out of the plane of the paper during the rotation. The two drawings do not represent the same molecule.

Q-21 One additional type of valid manipulation does not fit in the category just discussed. A tetrahedral grouping can be rotated around the axis defined by the bond between the central carbon and any vertex. Is the following manipulation valid?

CHO, H, OH, CH_2OH → CHO, HO, CH_2OH, H

A-21 Yes, it is proper. The drawing has been rotated about the bond between the CHO group and the central carbon.

CHO, H, OH, CH_2OH → CHO, HO, CH_2OH, H

The two drawings represent the same molecule.

Q-22 Can any more equivalent projections be obtained while keeping positions of the central carbon and the CHO group fixed?

A-22 Yes, as follows:

CHO, HO, CH_2OH, H → CHO, $HOCH_2$, H, OH

S-4 One of the most useful types of drawings is the "end-on" (Newman) projection. The molecule is shown as if one were looking down the carbon-carbon bond. The bonds from the front carbon intersect in the center of the circle. Bonds to the back carbon are drawn only to the edge of the circle.

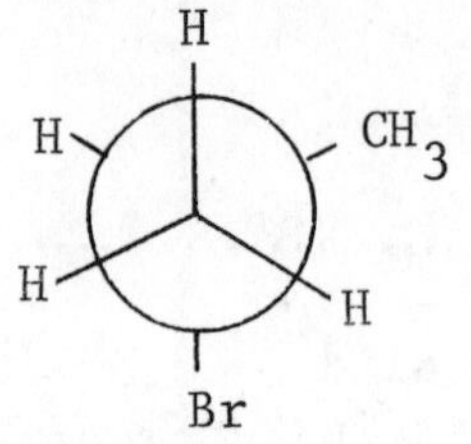

"end-on" (Newman) projection

Q-23 Draw a Newman projection of ethane.

A-23

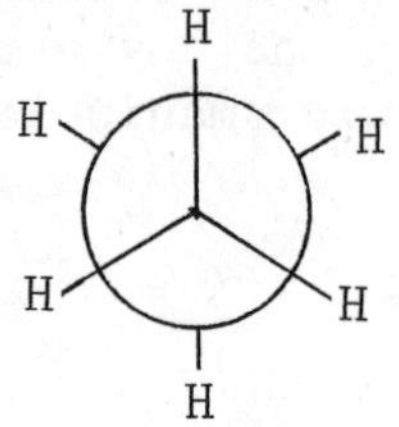

Q-24 Draw a Newman projection of ethanol.

A-24

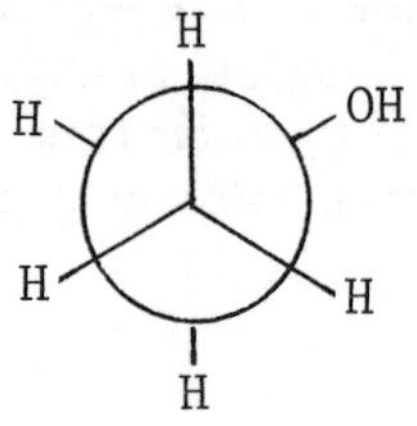

S-5 Still another important drawing convention is the "sawhorse". The molecule is pictured as it would appear to an observer who is located above and to the right of the molecule.

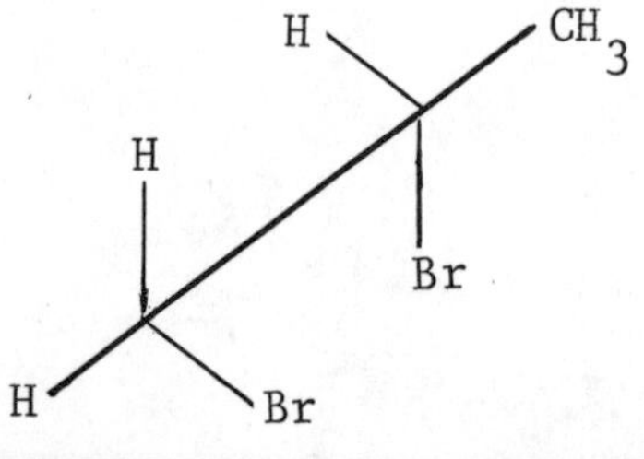

Q-25 Draw a sawhorse projection of ethane.

A-25

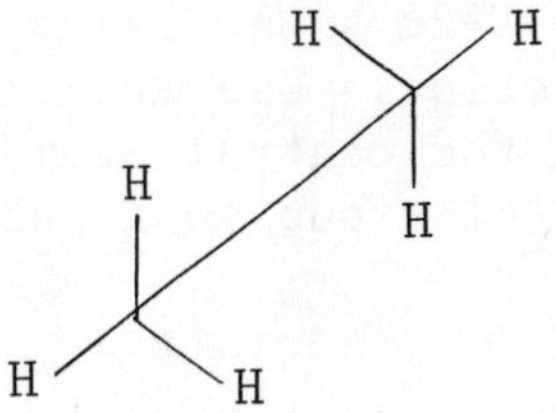

Q-26 Draw a sawhorse projection of chloroethane.

A-26

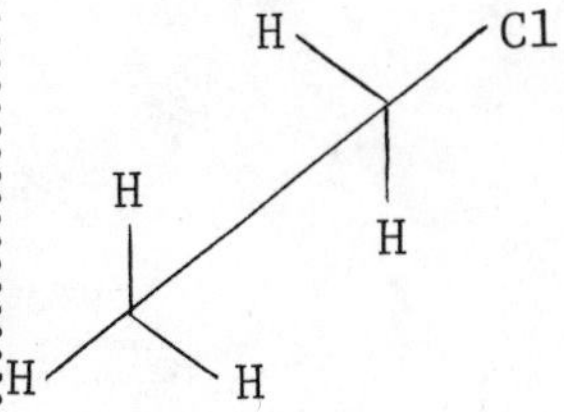

or

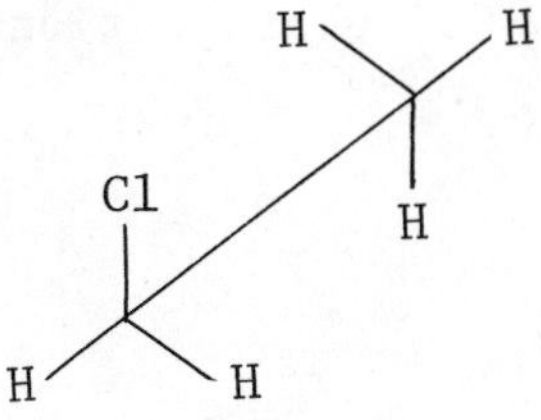

(1) (2)

Note that there are two other possible positions for the Cl group in each.

Q-27 Draw a Newman projection of chloroethane.

A-27 Note that (3) results from (1) and (4) from (2) just by changing the position of the observer slightly.

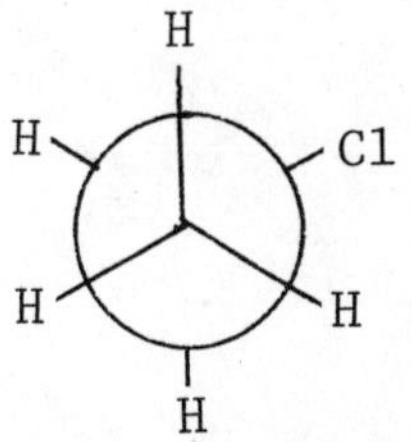

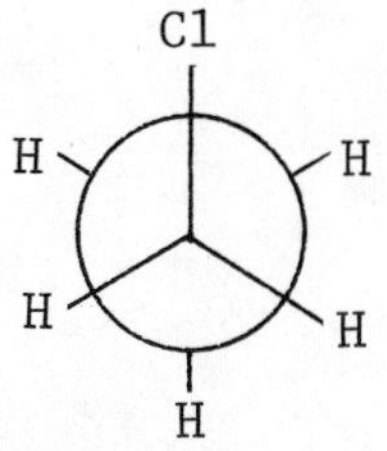

(3) (4)

Q-28 Convert the Newman projection below to a sawhorse projection.

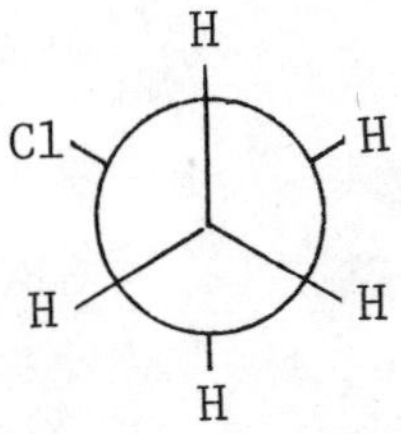

A-28

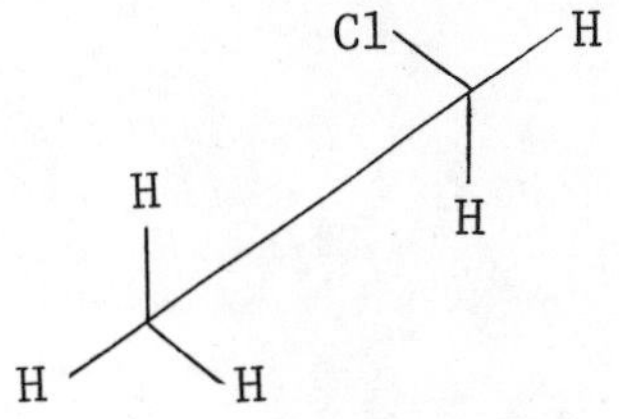

Q-29 Convert the sawhorse projection in A-28 to a flying wedge projection. Use the front carbon as the central atom and the CH_2Cl group as a substituent.

A-29

Q-30 Convert the drawing below directly into a flying wedge projection.

A-30

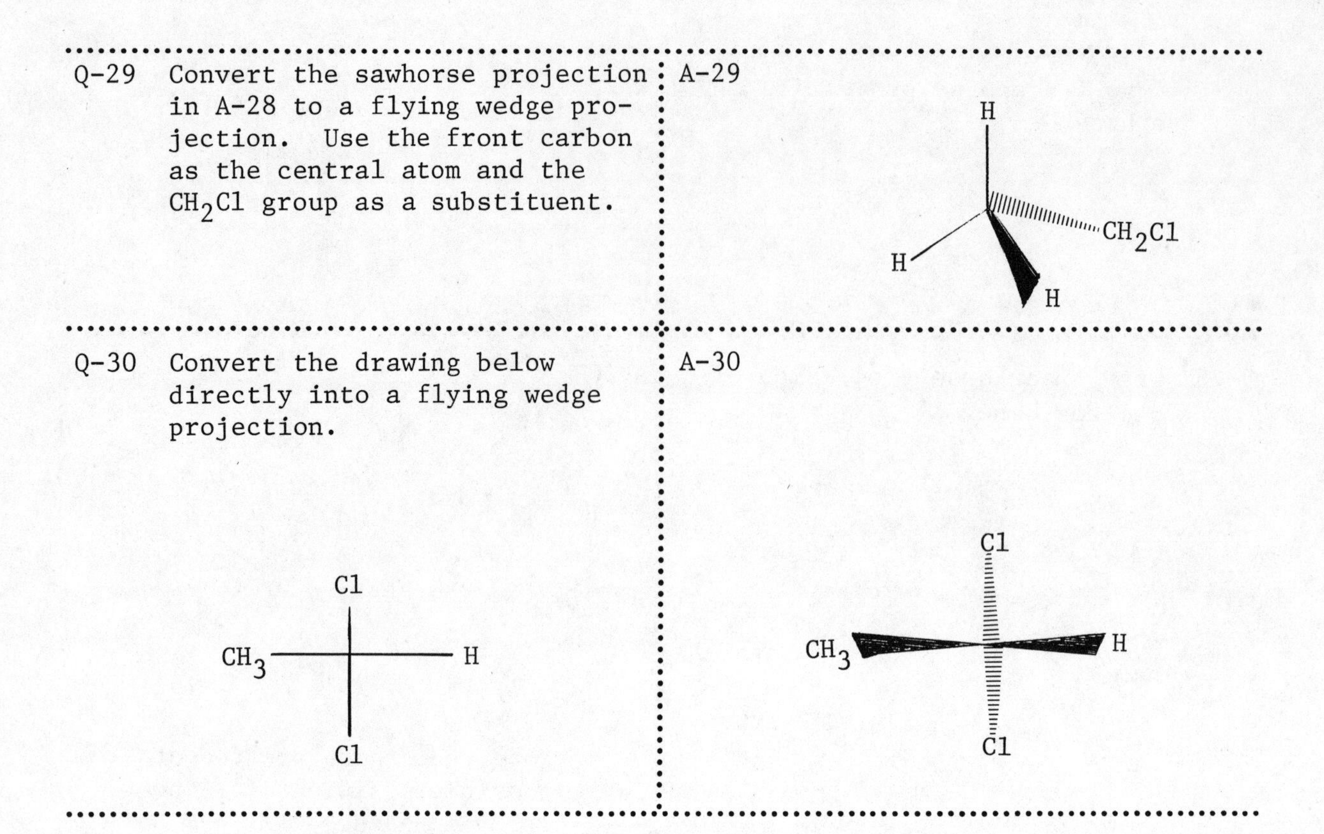

Section 2: Molecular Sources of Stereochemistry

Isomers:

S-1 A single molecular formula may represent several compounds which possess different physical and chemical properties. Different compounds with the same molecular formula are called isomers. The three basic kinds of isomers are position isomers, functional group isomers and stereoisomers.

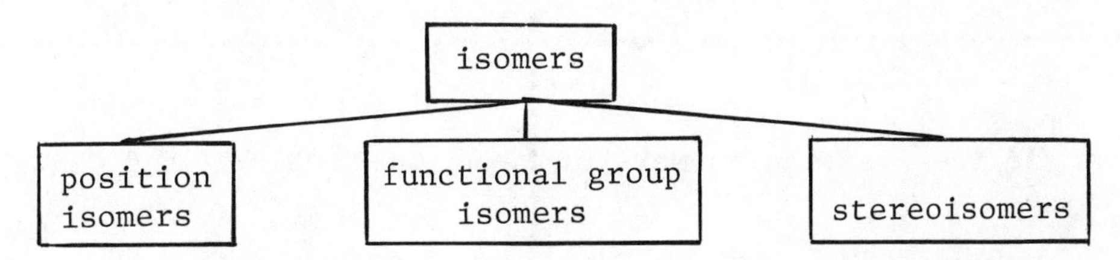

Position isomers have the same arrangement of atoms in the skeleton of the molecule but differ in the position of a substituent atom or group.

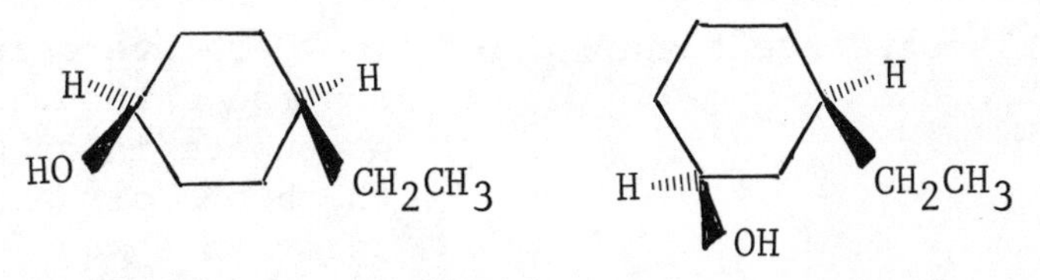

position isomers

Functional group isomers have different substituent groups.

functional group isomers

Stereoisomers have the same carbon skeleton and substituent groups but differ in the way some of the bonds are oriented in space.

stereoisomers

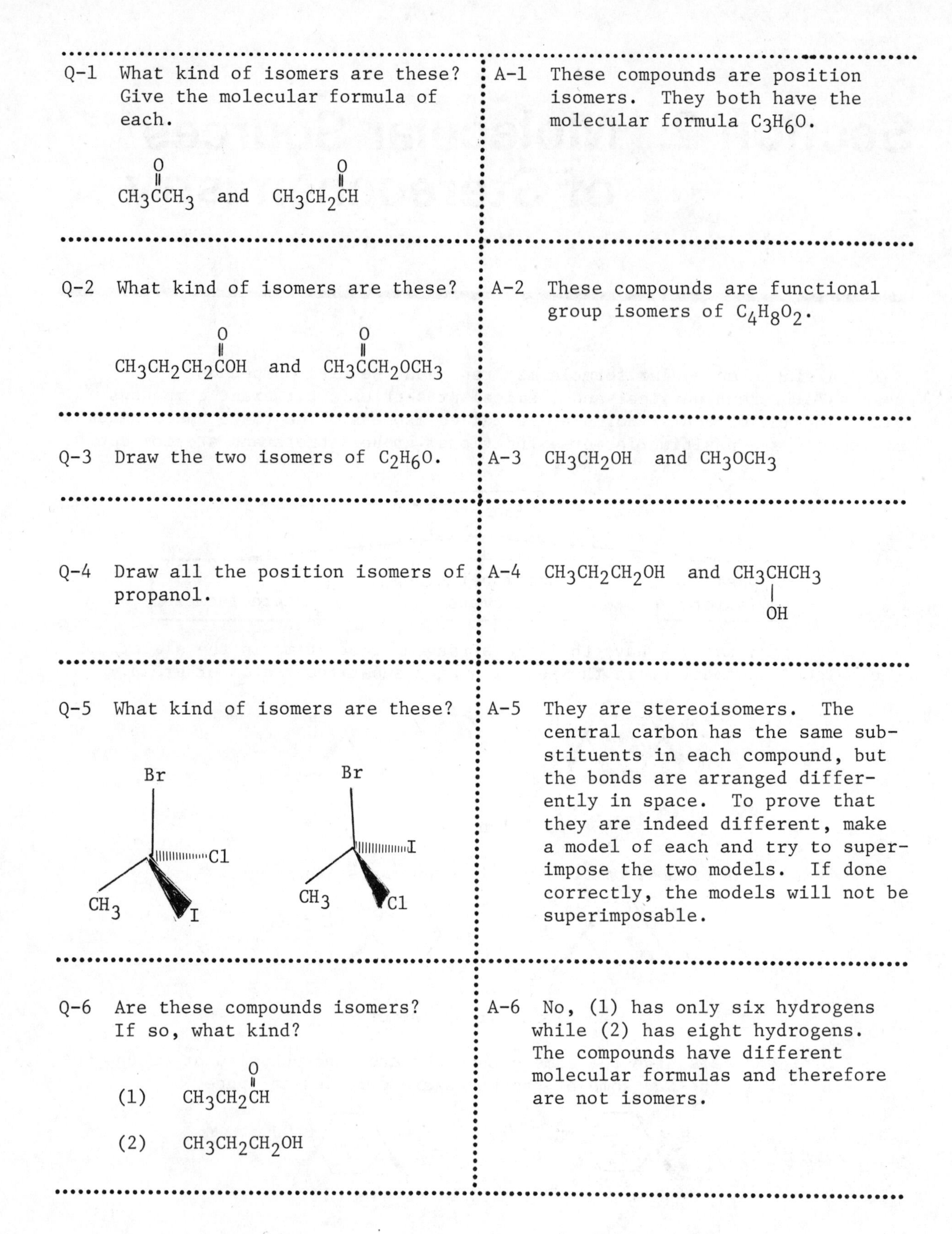

Q-1 What kind of isomers are these? Give the molecular formula of each.

$CH_3\overset{O}{\overset{\|}{C}}CH_3$ and $CH_3CH_2\overset{O}{\overset{\|}{C}}H$

A-1 These compounds are position isomers. They both have the molecular formula C_3H_6O.

Q-2 What kind of isomers are these?

$CH_3CH_2CH_2\overset{O}{\overset{\|}{C}}OH$ and $CH_3\overset{O}{\overset{\|}{C}}CH_2OCH_3$

A-2 These compounds are functional group isomers of $C_4H_8O_2$.

Q-3 Draw the two isomers of C_2H_6O.

A-3 CH_3CH_2OH and CH_3OCH_3

Q-4 Draw all the position isomers of propanol.

A-4 $CH_3CH_2CH_2OH$ and $CH_3\underset{OH}{\underset{|}{C}}HCH_3$

Q-5 What kind of isomers are these?

A-5 They are stereoisomers. The central carbon has the same substituents in each compound, but the bonds are arranged differently in space. To prove that they are indeed different, make a model of each and try to superimpose the two models. If done correctly, the models will not be superimposable.

Q-6 Are these compounds isomers? If so, what kind?

(1) $CH_3CH_2\overset{O}{\overset{\|}{C}}H$

(2) $CH_3CH_2CH_2OH$

A-6 No, (1) has only six hydrogens while (2) has eight hydrogens. The compounds have different molecular formulas and therefore are not isomers.

Q-7 Are the compounds shown below stereoisomers?

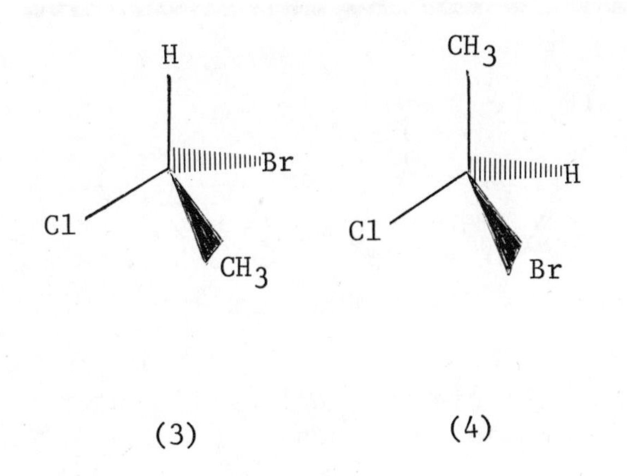

A-7 No. They are the same compound. The two drawings simply represent different perspectives. In determining whether two compounds are stereoisomers, care should be taken to exclude the possibility that the compounds appear different simply because they are being viewed from different directions.

Rotation of (3) as shown converts it into (4).

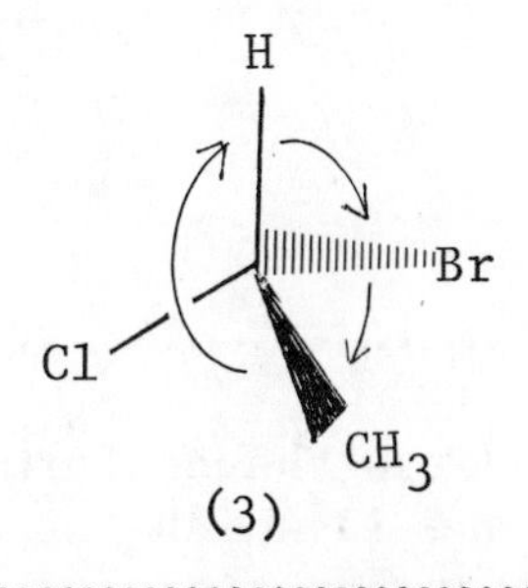

Q-8 Are the compounds shown below stereoisomers?

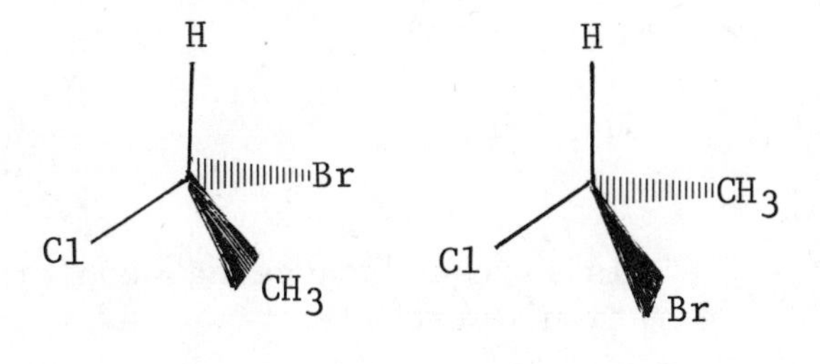

A-8 Yes. There is no way the two structures can be superimposed.

Q-9 Which of the compounds shown below are stereoisomers?

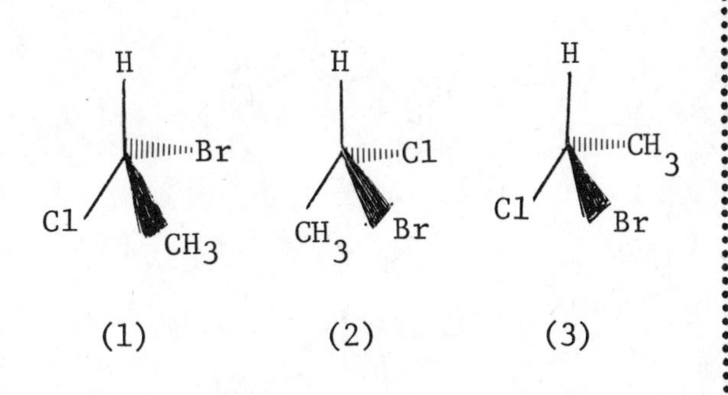

A-9 Compounds (2) and (3) are stereoisomers, as are (1) and (3). Actually (1) and (2) are different views of the same molecule.

Conformations:

S-2 Stereoisomers which can be interconverted by rotations about single bonds are called conformers or conformational forms.

Q-10 Ethane can adopt many rotational conformations. Draw the form in which the carbon-hydrogen bonds of one carbon are superimposed on the analogous bonds of the other carbon atom.

A-10

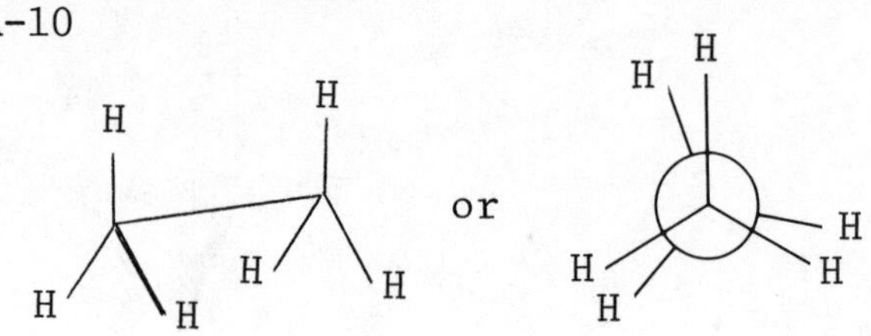

H H H H H H or H H H H H H

This is the "eclipsed" conformation of ethane.

Q-11 Now draw the conformation of ethane in which the C-H bonds on one carbon atom are as far as possible from the C-H bonds on the other carbon atom.

A-11

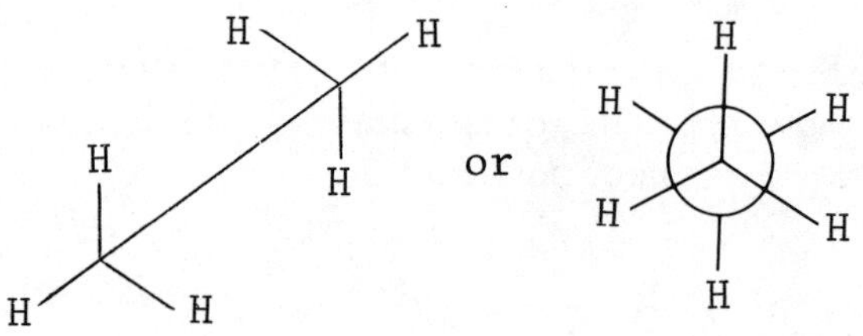

H H H H H H or H H H H H H

This is the "staggered" conformation of ethane.

Q-12 Draw the eclipsed conformation of propane using a Newman projection.

A-12

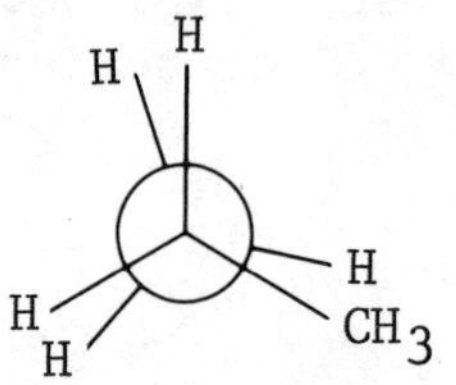

H H H CH_3 H H

or

H H CH_3 H H H

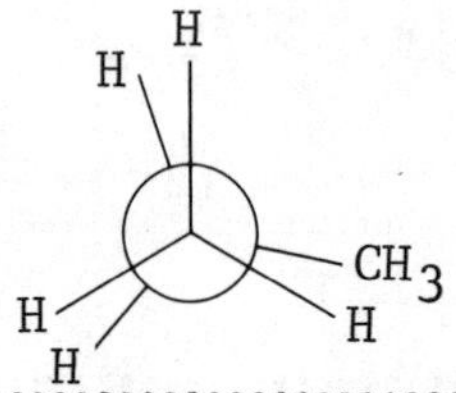

Q-13 Draw the staggered conformation of propane using a Newman projection.

A-13

Q-14 Would you expect the staggered or eclipsed conformation to be more stable? Why?

A-14 The staggered conformation is more stable for propane and for most other molecules. In the eclipsed conformation, the eclipsed groups crowd each other, leading to strain. In the staggered conformation the groups are as far apart as possible, thus minimizing the strain. In other words, the forces which act between non-bonded atoms can be important in determining the relative spatial locations of the atoms.

S-3 In a molecule such as *n*-butane there is more than one staggered conformation:

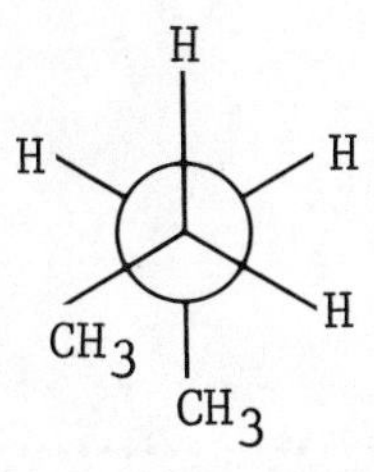

anti-conformation

gauche-conformations

Q-15 Which conformation is likely to be more stable, the *anti*- or the *gauche*? Why?

A-15 The *anti*- form is more stable. In the *gauche*- form the methyl groups are close together, producing some steric strain.

Q-16 Which of all the possible conformations of *n*-butane is likely to be least stable? Why?

A-16 The form in which the two methyl groups are eclipsed is the least stable. This conformer has the maximum amount

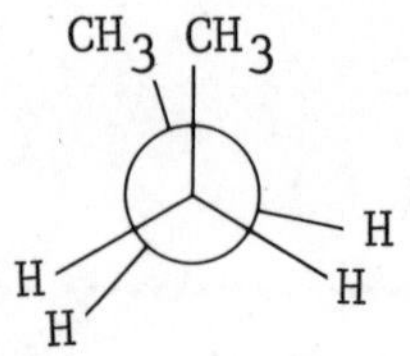

CH_3 CH_3
H H
H H

of strain due to a methyl-methyl steric interaction.

Q-17 Draw the *anti* conformer of 1,2-dichloroethane.

A-17

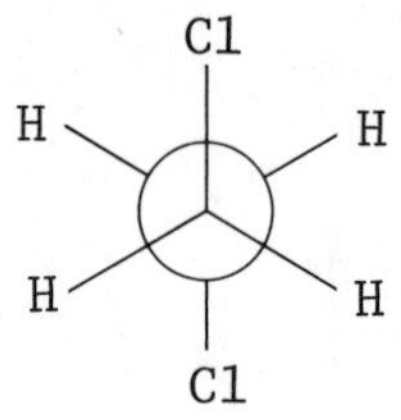

Cl
H H
H H
Cl

Q-18 Draw the two *gauche* conformers of 1,2-dichloroethane.

A-18

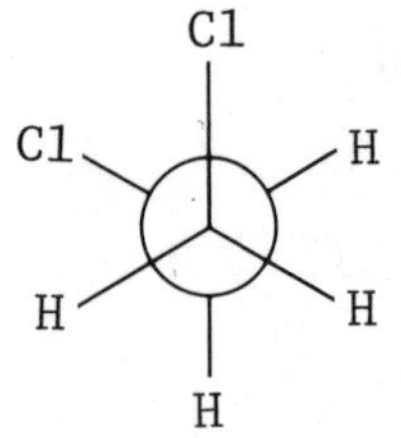

Cl
Cl H
H H
H

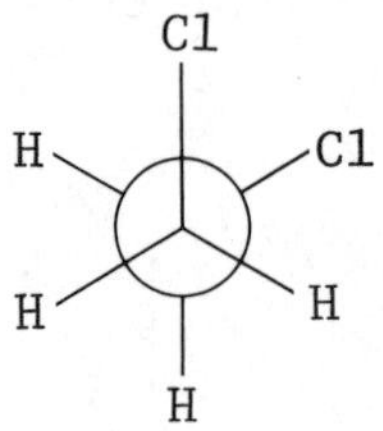

Cl
H Cl
H H
H

Q-19 Draw the two staggered conformations of $CHCl_2$-$CHCl_2$. Which is more stable? (Clue: Cl is bigger than H.)

A-19

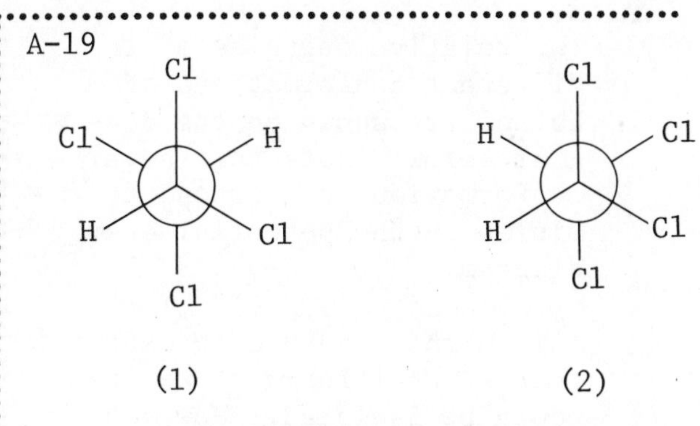

Conformation (1) is more stable since it has only two Cl-Cl strain-producing *gauche* interactions, while conformation (2) has three Cl-Cl *gauche* interactions.

Isolation of Stereoisomers:

S-4 Stereoisomers can be isolated whenever they are constrained from interconverting readily. This is the case whenever they are separated by a significant energy barrier (roughly 80 kJ/mol or more at room temperature).*

Q-20 Why are the atoms in a molecule not free to adopt any spatial arrangement?

A-20 The bonds linking the atoms place an important constraint on the freedom and motion of the atoms.

* Note: 1 cal (calorie) = 4.18 J (joule) and 1 kcal = 4.18 kJ. A range of 75 - 85 kJ/mole therefore is approximately equivalent to 18 - 20 kcal/mole.

Q-21 The relative energies of the different conformations of *n*-butane are shown on the diagram below. Note that several conformations correspond to minima in the potential energy diagram.

This might lead one to expect that several forms of *n*-butane could be isolated. However, this is not the case. Explain why not.

A-21 The energy barriers between the several relatively stable conformations are low, resulting in their rapid interconversion at room temperature. *n*-Butane exists as an equilibrium mixture of these different conformations. Some of these conformations can interconvert at rates up to a trillion times a second.

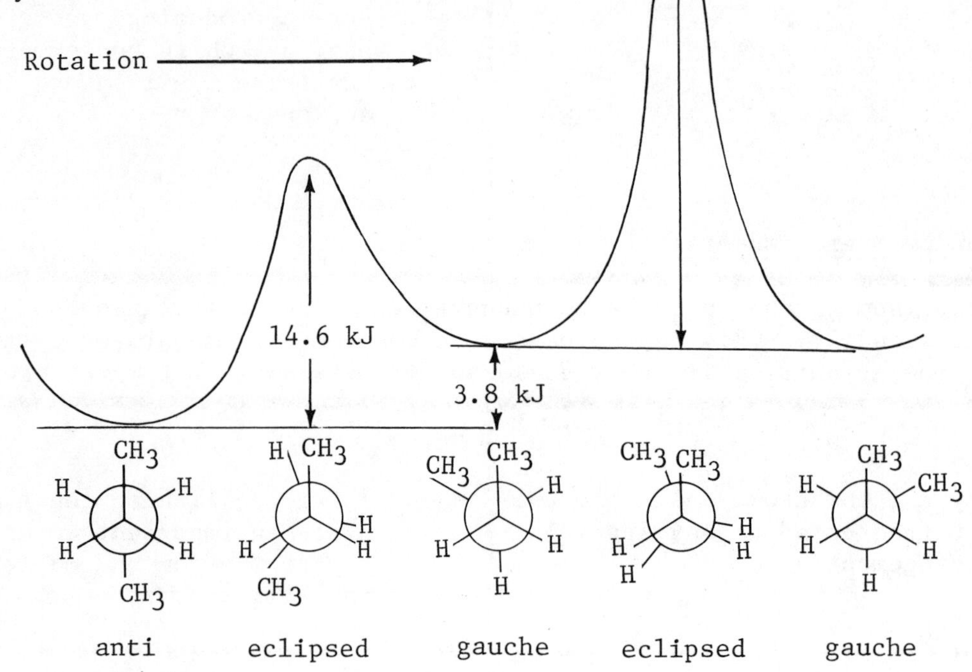

Q-22 Are the different conformational forms of *n*-butane stereoisomers?

A-22 Yes, since the spatial arrangement of the atoms is different in each conformation.

Q-23 How might some additional constraints, beyond the single bonds already present, be imposed on the atoms of *n*-butane to further restrict their movement?

A-23 The addition of more chemical bonds would provide further constraints. For instance, a single bond could be added to afford cyclobutane. A double bond could be added to form a butene

$$\begin{array}{c} CH_2-CH_2 \\ | \quad\quad | \\ CH_2-CH_2 \end{array} \qquad \begin{array}{c} \quad\quad\quad CH_3 \\ CH=CH \\ CH_3 \quad\quad\quad \end{array}$$

cyclobutane   *trans*-2-butene

Q-24 In which of the two substances shown in the previous question does the additional bond give rise to a compound which has a stereoisomer which can be isolated?

A-24 *Trans*-2-butene is a stereoisomer which can be isolated because it is separated by a significant energy barrier from *cis*-2-butene, which is a stereoisomer of *trans*-2-butene.

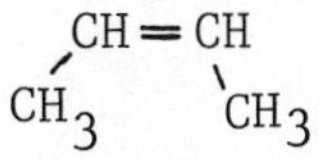

CH=CH
CH$_3$ CH$_3$

Q-25 Can you imagine any other source of restricted rotation in molecules besides additional chemical bonds?

A-25 Yes, crowding of atoms or groups. Rotation would be limited (perhaps even to the extent of allowing the stereoisomers to be isolated) by atoms getting in each other's way. (Atoms cannot occupy the same place at the same time.)

Q-26 The most stable conformation for biphenyl derivatives is the one in which the planes of the two aromatic rings are at a 90° angle to each other. Draw this conformation for biphenyl.

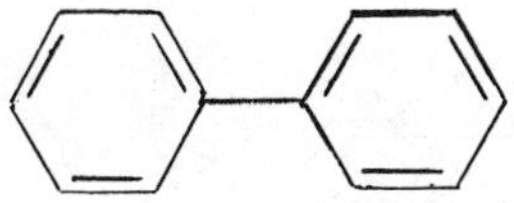

A-26

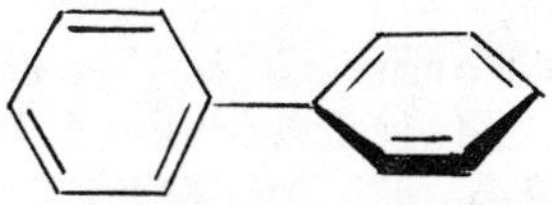

The left-hand ring is in the plane of the paper, the right-hand one at a right angle to the plane.

Q-27 Why do you suppose this conformation is the most stable one for biphenyl?

A-27 It minimizes steric crowding between the two rings and the H atoms on the rings. This crowding would be maximal when the rings are planar and minimal when they are perpendicular.

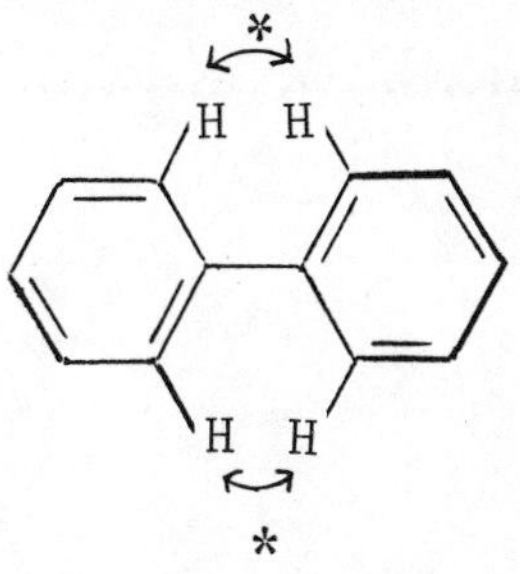

H H
H H

* steric crowding

Q-28 Draw the most stable conformation of 2,2'-dichlorobiphenyl.

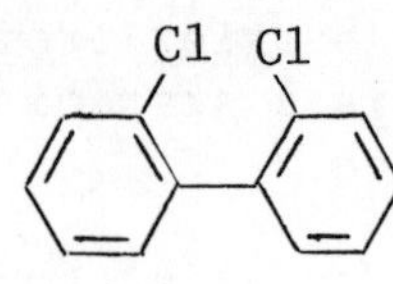

A-28 There are two more stable conformations as shown below.

Q-29 Although there are two stable conformations for 2,2'-dichlorobiphenyl only one form of the compound has ever been isolated. Explain.

A-29 The two conformations can interconvert rapidly by rotation around the single bond joining the two rings. As a result the form isolated is a mixture of the two conformations. The steric crowding in the planar conformation through which the two conformers must rotate to interconvert is not severe enough to prevent interconversion at normal temperatures.

Q-30 Two forms of 2,2'-dichloro-5,5'-diiodobiphenyl have been isolated. Explain.

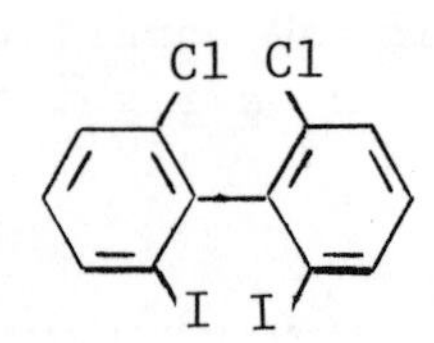

A-30 In this case there is sufficient restriction to rotation to prevent the interconversion of the two conformations.

(1) (2)

The interconversion is difficult because the bulky *ortho* groups (the chlorine and iodine atoms) must be forced past each other.

Cyclohexane Conformations:

S-5 Although rotation around a single bond is normally relatively easy, formation of a ring greatly restricts the degree of rotation around the single bonds which make up the ring. As a result the number of possible spatial arrangements of the atoms (conformations) is severely restricted. Several important conformations of cyclohexane are shown below.

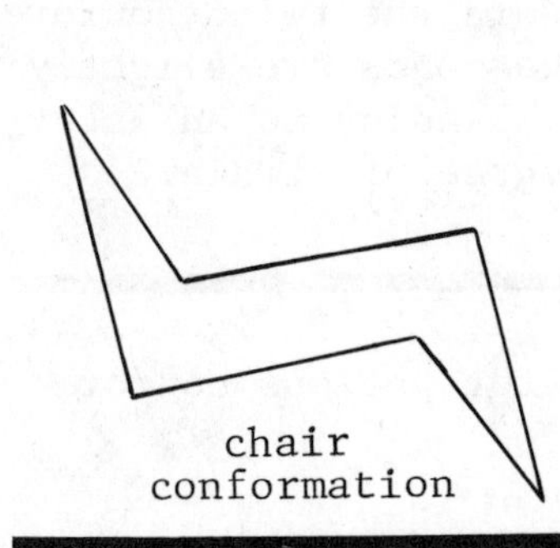

chair conformation

boat conformation

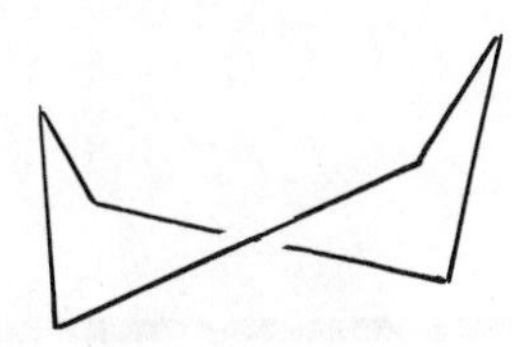

flexible (twist) conformation

Q-31 Which conformation of cyclohexane is represented in the drawing below. (Hint: Look at a model.) What type of projection is used?

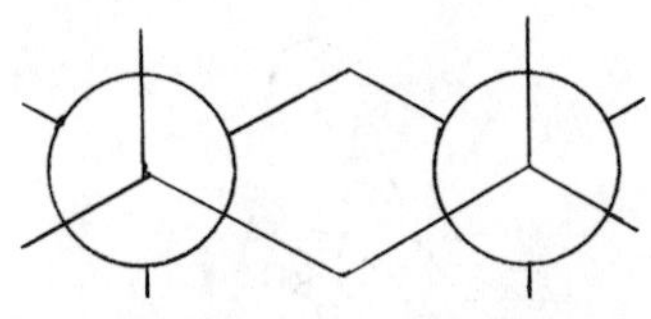

A-31 The drawing shows a Newman projection of the chair conformation. Note that the bonds are perfectly staggered.

Q-32 Draw a Newman projection of the boat conformer of cyclohexane. (Use a model.)

(conformer = conformation)

A-32

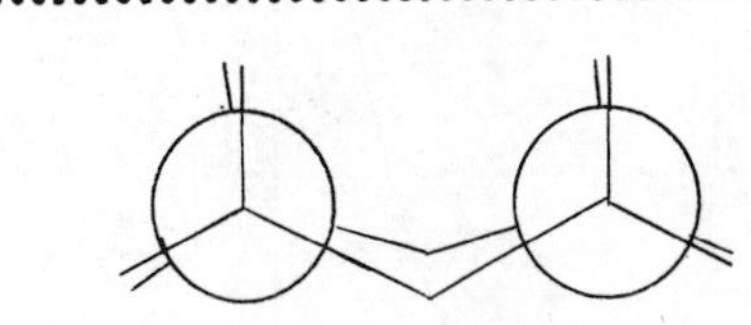

Newman projection of the boat conformation of cyclohexane. The bonds are exactly eclipsed.

Q-33 Would you expect the chair or boat conformation to be more stable? Why?

A-33 The chair conformation should be more stable than the boat. The boat conformation is destabilized by twisting strain resulting from the eclipsing bonds. The chair conformation is relatively free of such strain.

Q-34 In the flexible (twist) conformation, the bonds are slightly staggered. Predict the stability of this conformation relative to the boat and chair conformations. Explain briefly. Make a model.

A-34 The predicted order of stability is chair > twist > boat.

In the chair form strain is minimized by the perfect staggering of bonds. The strain is highest in the boat form, where the bonds are completely eclipsed. In the twist conformation, the bonds are slightly staggered, leading to an intermediate degree of strain.

S-6 This energy diagram gives the relative stabilities for various conformations of cyclohexane:

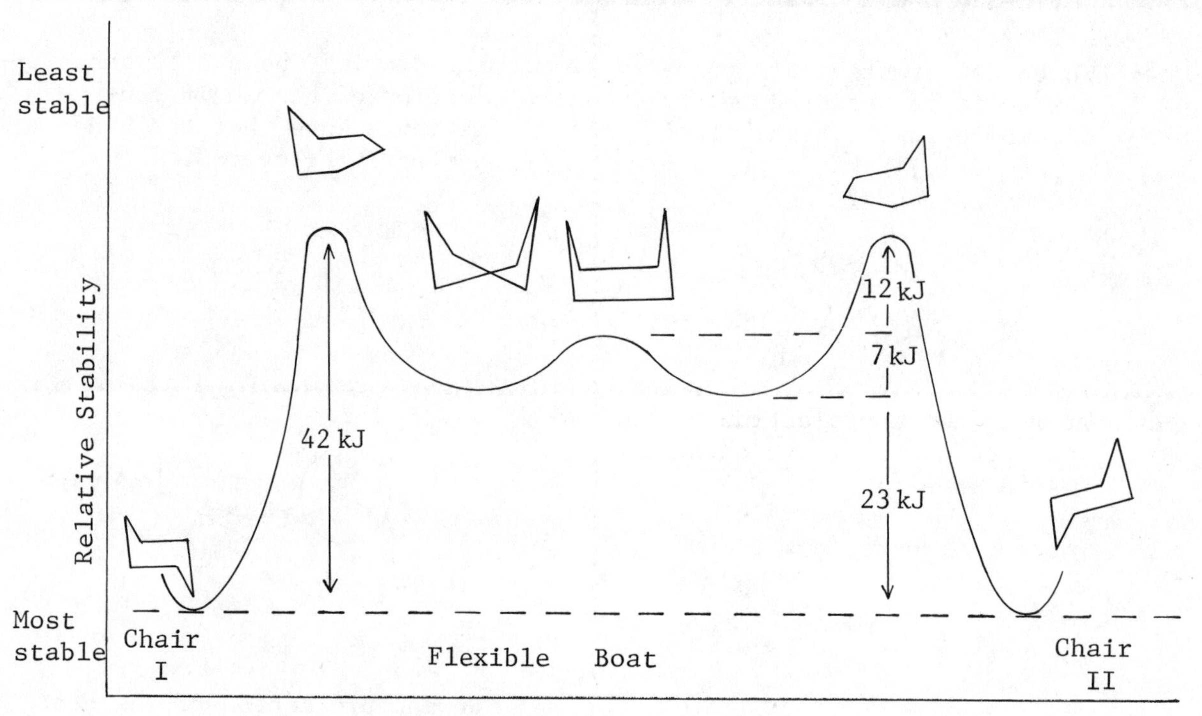

Note that chair form I can be converted to chair form II through several conformations of higher energy. (kJ units are defined on page 21.)

Q-35 Starting with a model of chair form (I), perform the following sequence of conversions:

chair form I → flexible form → boat form → chair form II

Are the molecules represented by chair forms I and II any different?

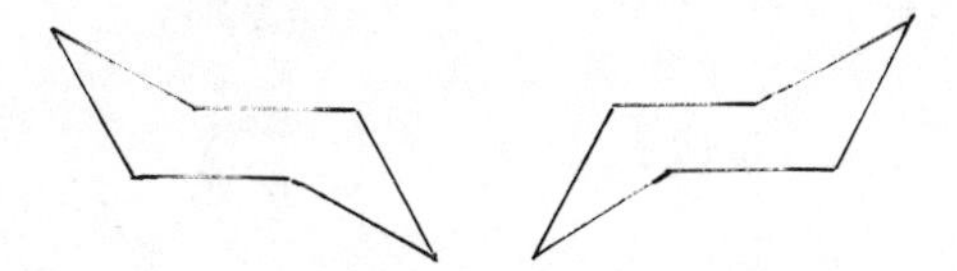

chair form I   chair form II

A-35 No, one can superimpose one chair form on the other after rotating one of them 180°, or by tipping one of them in the plane of the paper.

Q-36 Room temperature allows molecules to clear a 40 kJ/mole (10 kcal/mole) energy barrier rather easily. List the following in order of increasing probability of occurrence:

boat form, flexible form, chair form I, chair form II

A-36
1. boat
2. flexible
3. chair I = chair II

(increasing probability of occurrence ↓)

Those conformers with the least energy are the most stable and therefore the most probable. About one molecule in a thousand will be in the boat form at room temperature.

Q-37 Steric strain often arises when groups are forced close together. Is there any strain in the boat form in addition to the strain from eclipsing?

A-37 Yes. The hydrogens at either "end" are brought close together, resulting in additional strain.

interaction between hydrogen atoms

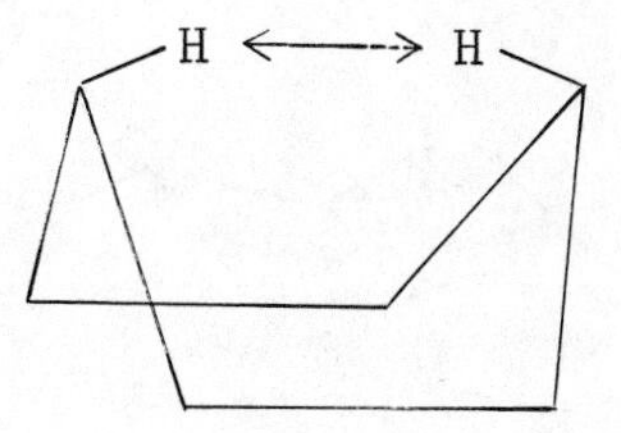

S-7 The hydrogen atoms attached to the cyclohexane ring can be oriented either approximately in the plane of the ring (equatorial or e) or vertically (axial or a). Note that all axial bonds are parallel to each other. Every equatorial bond is parallel to the two ring bonds which are one bond removed from the equatorial bond.

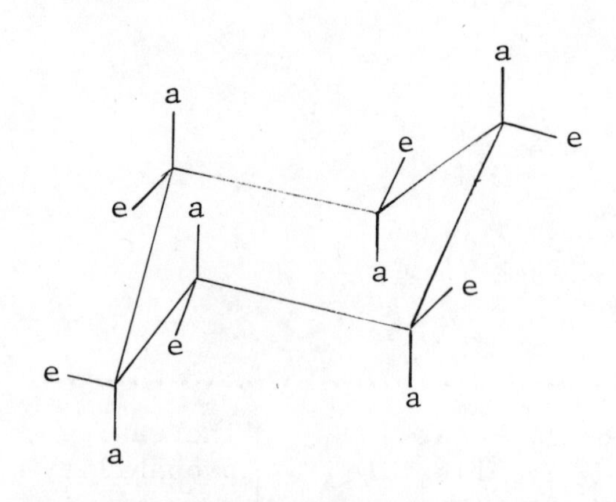

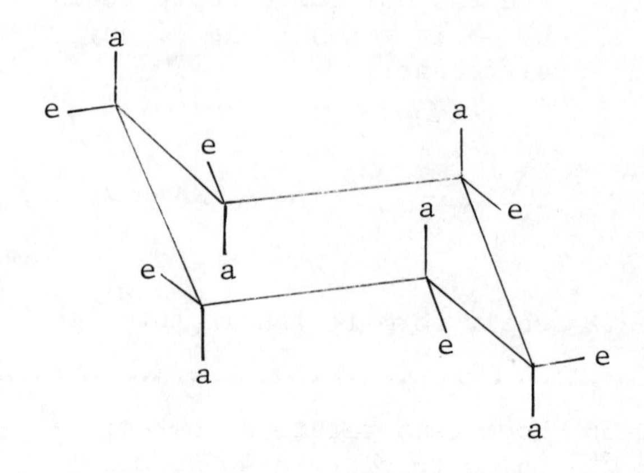

Q-38 Draw the axial bonds on this chair form.

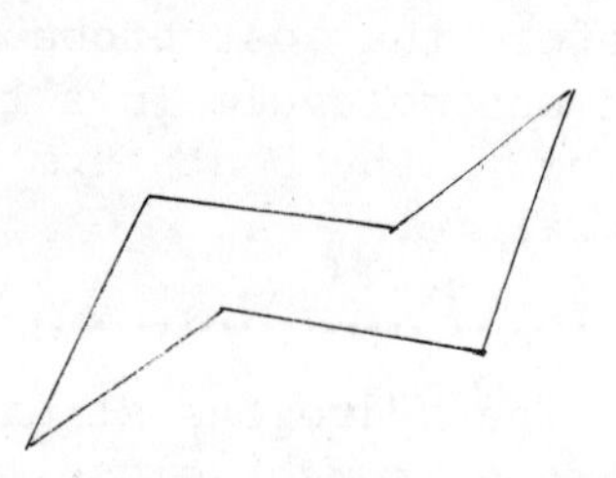

A-38

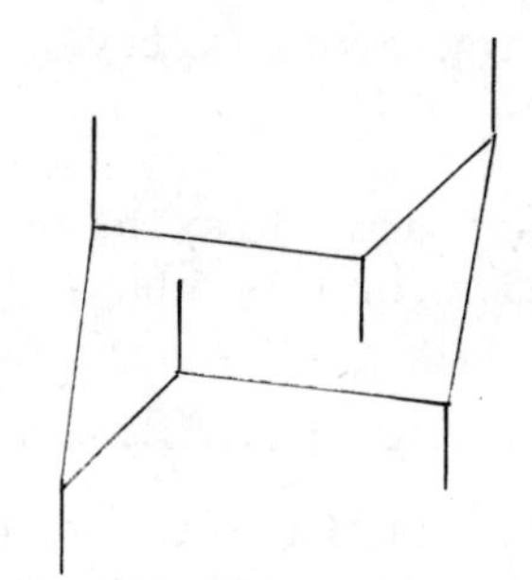

Q-39 Darken the two ring bonds which are parallel to the equatorial bond which is shown below. Also, add the rest of the equatorial bonds.

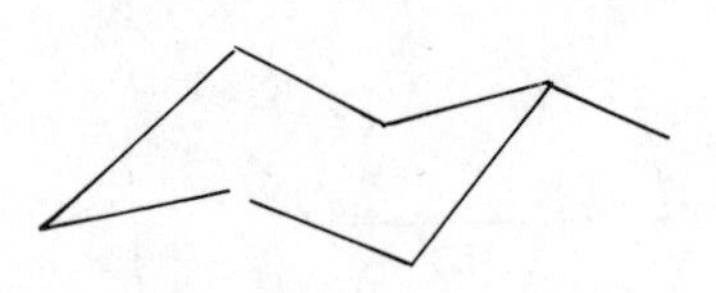

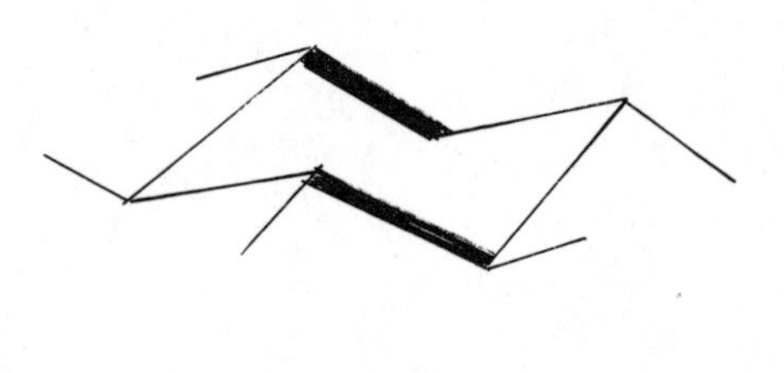

Q-40 Draw a Newman projection of this structure, using the 1, 2 and 5, 4 carbon-carbon bonds as the axes.

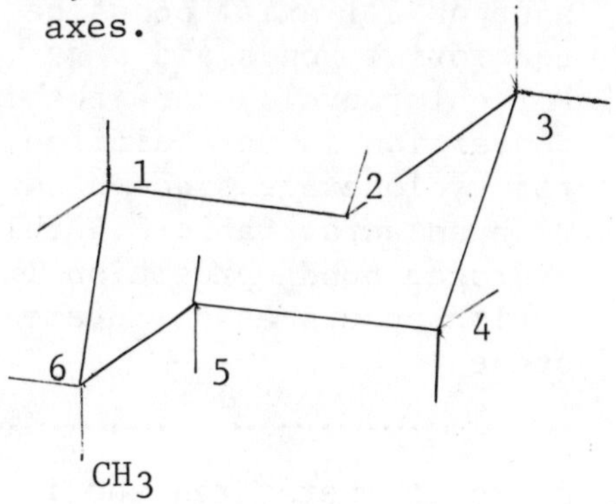

1 2 3 4 5 6 CH_3

A-40

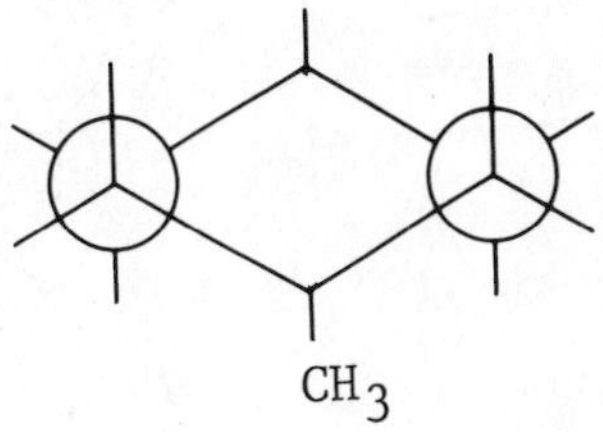

CH_3

Q-41 Draw the Newman projection of 1,2-dibromocyclohexane looking down the 1, 2 and 5, 4 bonds. Place both bromo groups in axial positions.

A-41

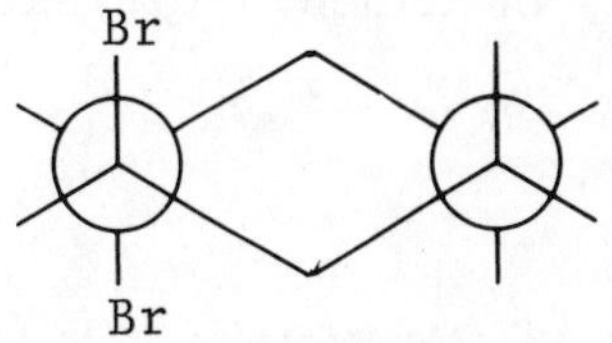

Br

Br

Q-42 Draw the Newman projection of 1,2-dihydroxycyclohexane looking down the 1, 2 and 5, 4 bonds. Place the one hydroxy group in an axial position and the other in an equatorial position.

A-42

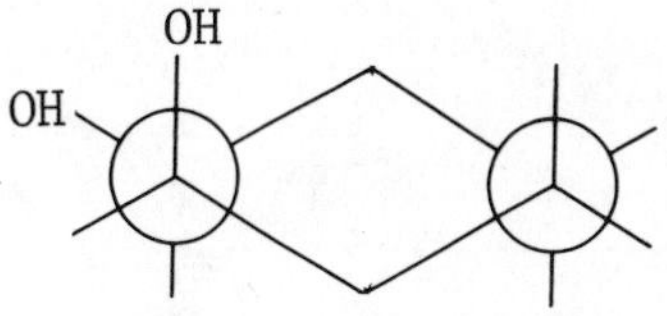

OH OH

or

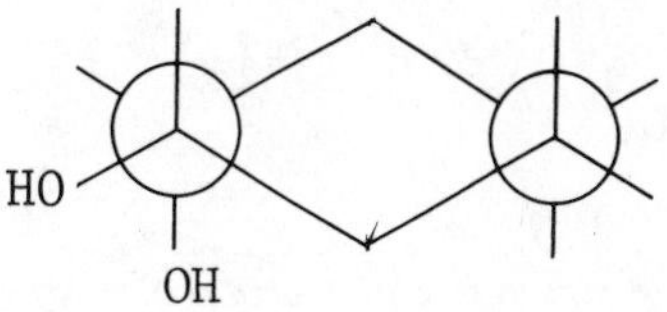

HO OH

Q-43 Draw the Newman projection of 1,2-dimethylcyclohexane looking down the 1, 2 and 5, 4 bonds. Place both methyl groups in equatorial positions.

A-43

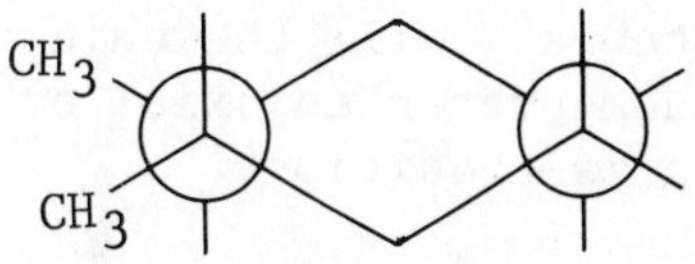

CH_3 CH_3

Q-44 Are chair forms I and II equivalent?

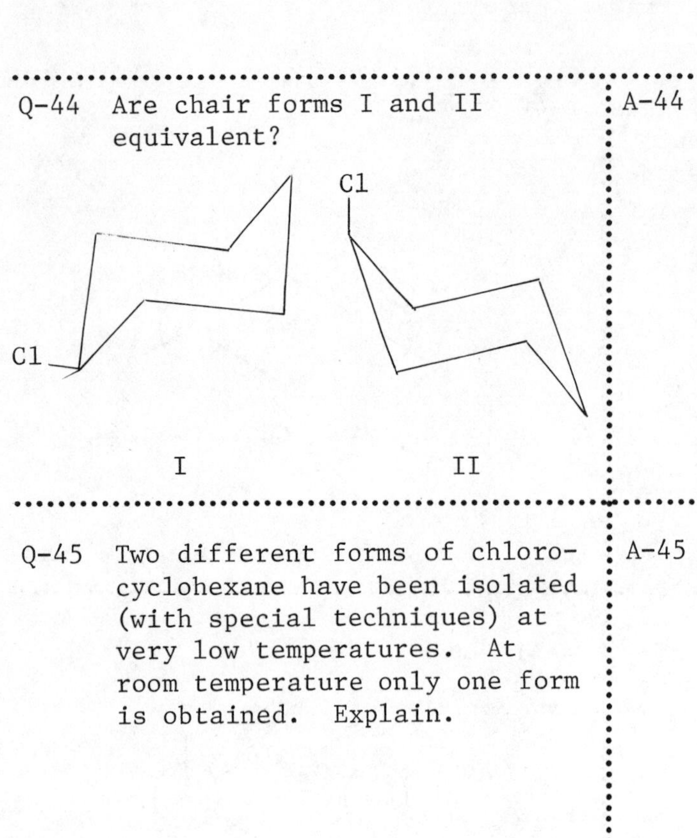

A-44 No. However, these conformations can interconvert by "flip-flopping." Note that when this happens all axial bonds become equatorial bonds and *vice versa*. For chlorocyclohexane the interconversion is more difficult than for cyclohexane since a boat form with an unfavorable 1,4 chlorine-hydrogen bond opposition interaction occurs as an intermediate stage.

Q-45 Two different forms of chlorocyclohexane have been isolated (with special techniques) at very low temperatures. At room temperature only one form is obtained. Explain.

A-45 At low temperatures the interconversion of the two chair forms is slow and each one can be obtained separately since they have different physical properties. At room temperature, however, interconversion is rapid. The chlorocyclohexane obtained is an equilibrium mixture of the two conformations.

Q-46 Carefully examine a model of II (Q-44). Where is steric crowding most likely to occur?

A-46 Between the chlorine atom and the two hydrogen atoms on the same side of the ring.

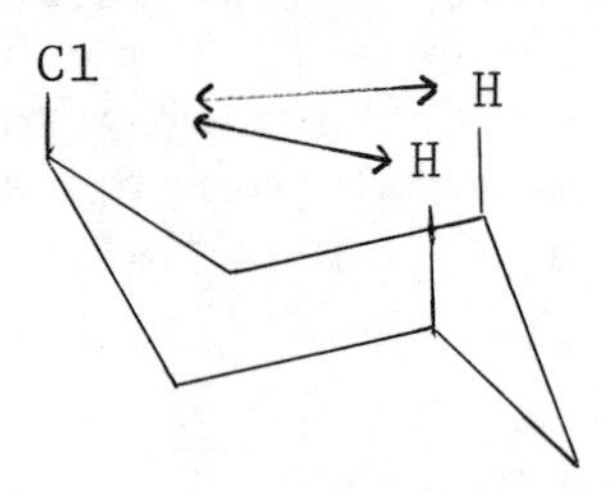

This is called 1,3 diaxial interaction or 1,3 transannular (across the ring) strain.

Q-47 Which chair conformation of chlorocyclohexane is more stable? Will the chlorine atom prefer an axial or equatorial position?

A-47 Conformation I, with the chlorine atom group in an equatorial position, is less crowded and therefore more stable. In II the chlorine atom interacts with the two axial hydrogens on the same side of the ring, causing strain in the molecule.

Q-48 Are the chair forms I and II stereoisomers?

A-48 Yes, since the directions of the bonds in space are different.

Q-49 Although placing a methyl, ethyl, or isopropyl group in an axial position instead of an equatorial position is about equally unfavorable for all three groups, placing a *t*-Bu group in an axial position is almost three times more unfavorable than these other groups. Suggest an explanation.

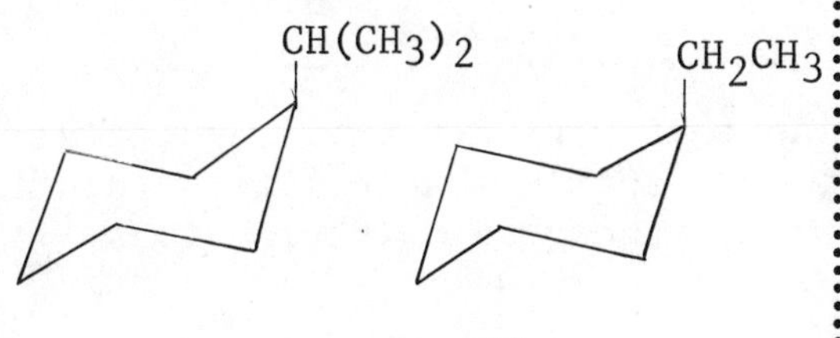

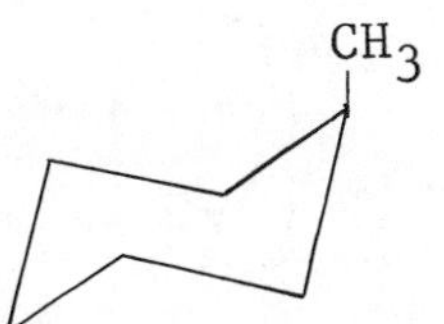

8 kJ/mol^{-1} unfavorable relative to equatorial

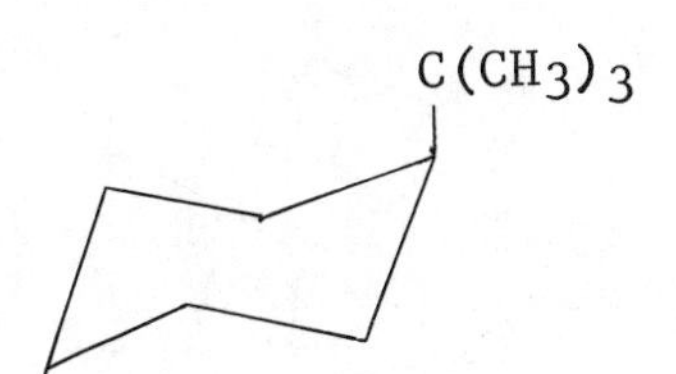

20 kJ mol^{-1} unfavorable relative to equatorial

A-49 An isopropyl or smaller group can turn in such a way that a hydrogen points back toward the ring. Thus essentially the same 1,3 interactions occur for each of these groups.

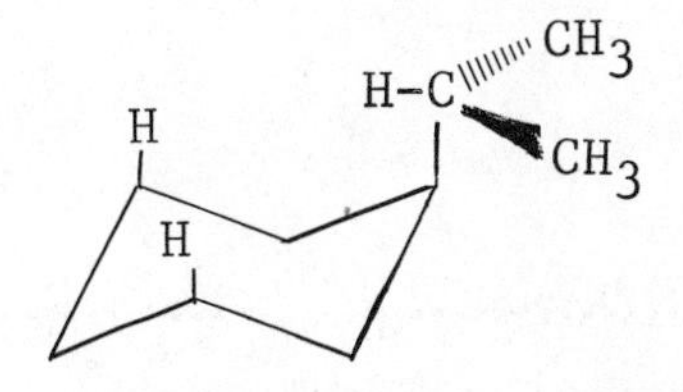

With an axial *t*-Bu group, however, one of the CH_3 groups must point in toward the ring, leading to a sizeable repulsion.

S-8 When a cyclohexane ring bears substituents on two or more ring carbon atoms, pairs of stable stereoisomers may exist. They are called *cis* and *trans* isomers by analogy with stereoisomers in double-bond compounds.

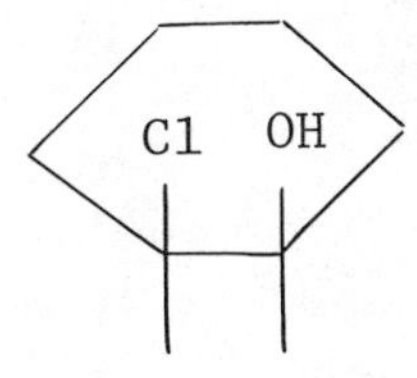

cis-2-chlorocyclohexanol

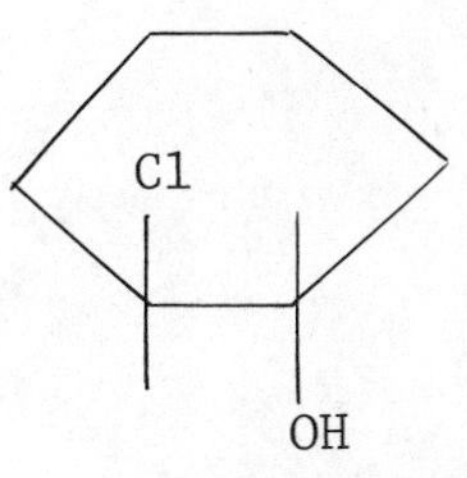

trans-2-chlorocyclohexanol

Note that in the *cis*-stereoisomer both substituents are on the same side of the ring; in the *trans*-compound they are on opposite sides.

Q-50 Draw the two chair conformations of *trans*-2-chlorocyclohexanol.

A-50

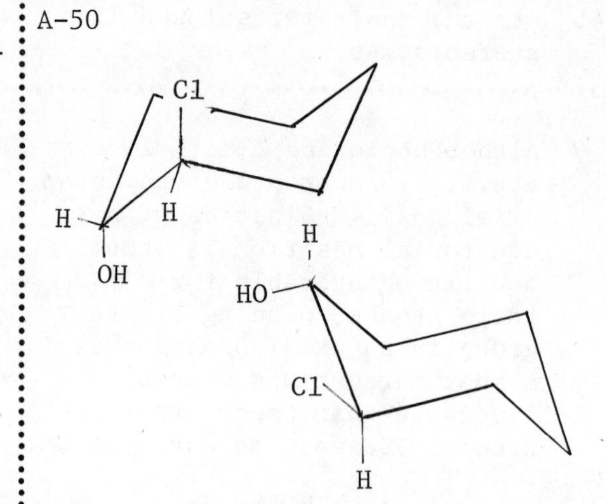

Q-51 The conformation with both substituents in the equatorial position has been found to be substantially more stable than the one with both in the axial position. How can this be explained?

A-51 Apparently the 1,3 diaxial interactions cause more strain than the *gauche* interactions.

1,3 diaxial HO H H

Cl H H

H

HO

Cl

gauche H

Q-52 Draw *cis*-2-chlorocyclohexanol in its chair conformation.

A-52

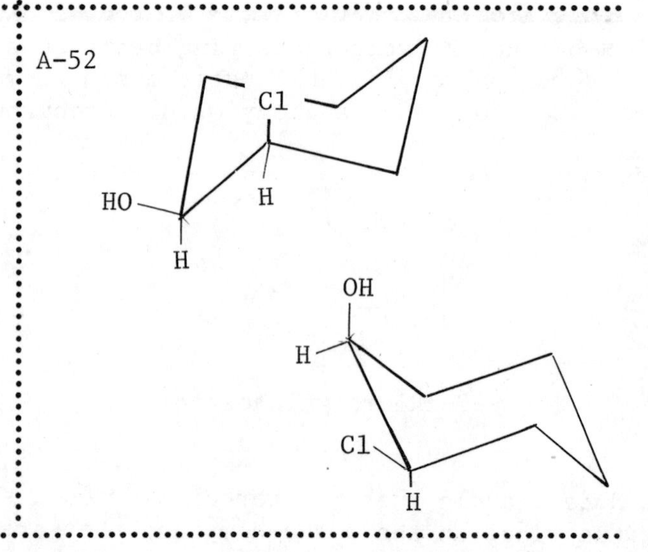

Cis - trans Isomerism:

S-9 In contrast to the relatively low energy barrier for rotation around a single bond, the energy required for rotation around a double bond is high. Such rotation therefore is very restricted under normal conditions so that stereoisomerism usually results.

Q-53 Are the compounds below stereoisomers?

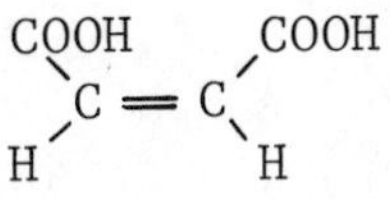

```
COOH     COOH    COOH      H
    \   /            \    /
     C = C            C = C
    /   \            /    \
   H     H          H      COOH
```

maleic acid 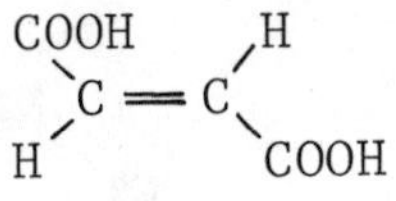   fumaric acid

A-53 Yes, they are stereoisomers. In each compound there is one carboxyl group and one hydrogen atom bonded to each carbon of the double bond. However, the bonds are arranged differently in space in the two compounds.

Q-54 Are these compounds stereoisomers?

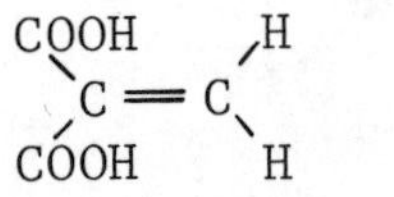

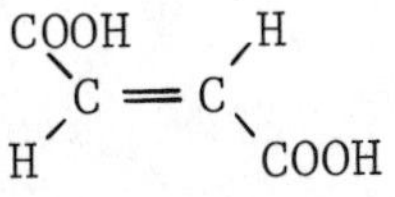

```
COOH     H     COOH      H
    \   /          \    /
     C = C          C = C
    /   \          /    \
COOH     H        H      COOH
```

A-54 No, the carboxyl groups are attached to the same carbon atom in one compound, but to different carbon atoms in the other.

Q-55 There are two stereoisomers which have the formula $C_2H_2Cl_2$. Draw them.

A-55 The pair of stereoisomers is

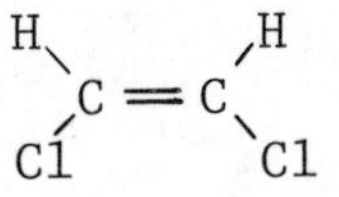

```
H       H            H       Cl
 \     /              \     /
  C = C       and      C = C
 /     \              /     \
Cl      Cl           Cl      H
```

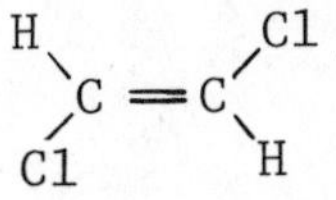

Note: Compound 1 is a position isomer of each of the above compounds.

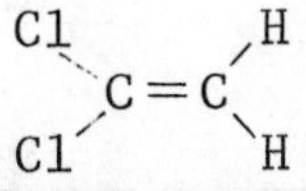

```
Cl      H
  \    /
   C = C       (1)
  /    \
Cl      H
```

S-10 Pairs of geometric stereoisomers involving carbon-carbon double bonds are designated *cis* or *trans*. These names refer to the position of substituents with respect to each other.

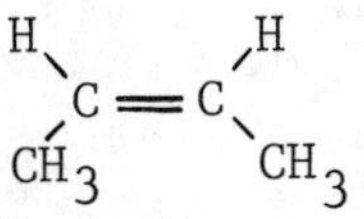

```
H       H              CH3      H
 \     /                  \    /
  C = C                    C = C
 /     \                  /    \
CH3     CH3              H      CH3
```

cis-2-butene 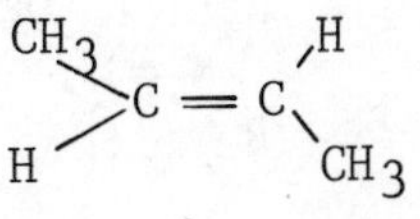   *trans*-2-butene

Note that in the *cis*-2-butene the methyl groups are on the same side of the double bond. In the *trans*-2-butene the methyl groups are on opposite sides of this double bond.

Q-56 Does the following compound show *cis-trans* isomerism? If so, draw the two isomers.

1,1-dichloroethylene

A-56 No, this structure has no stereoisomers.

```
Cl      H
  \    /
   C=C
  /    \
Cl      H
```

Q-57 Does the following compound show *cis-trans* isomerism? If so, draw the isomeric structures.

1,2-dichloroethylene

A-57 Yes, as shown.

```
H       Cl
  \    /
   C=C
  /    \
Cl      H
```

trans-

```
Cl      Cl
  \    /
   C=C
  /    \
H       H
```

cis-

Q-58 Draw the structure of the *cis*-stereoisomer of 2-pentene.

A-58

```
CH3      C2H5
  \      /
   C = C
  /      \
H         H
```

Q-59 Draw the structure of the *trans*-stereoisomer of 3-hexene.

A-59

```
C2H5      H
  \      /
   C = C
  /      \
H         C2H5
```

S-11 Stereoisomerism is also possible in compounds having carbon-nitrogen or nitrogen-nitrogen double bonds. For isomeric oximes the special terms *syn* and *anti* are used instead of *cis* and *trans*, respectively.

```
H       OH
  \    /
   C=N:
  /
R
```

a *syn*-oxime

(H and OH *cis*)

```
H     ..
  \    
   C=N
  /    \
R       OH
```

an *anti*-oxime

(H and OH *trans*)

Q-60 Does the following compound show *cis-trans* isomerism? If so, draw the isomeric structures.

$CH_3CH{=}NOH$

A-60 Yes, as shown.

$$\begin{array}{c} CH_3 \quad\quad\quad \\ \diagdown \quad\quad\quad\quad \\ C = \ddot{N} \quad \\ \diagup \quad\quad \diagdown \quad \\ H \quad\quad\quad\quad OH \end{array}$$

syn-

$$\begin{array}{c} CH_3 \quad\quad OH \\ \diagdown \quad\quad \diagup \\ C = \ddot{N} \quad \\ \diagup \quad\quad\quad\quad \\ H \quad\quad\quad\quad\quad \end{array}$$

anti-

Section 3: Nomenclature and Optical Activity in Stereochemistry

Elements and Operations:

S-1 Certain geometric *elements* and *operations* are used to help describe and analyze the stereochemical properties of molecules. The *elements* are points, lines and planes. A point element is always placed in the center of a molecule. A line element is drawn through the center of the molecule. A plane element is either drawn through the center of the molecule or placed completely outside the molecule. An *operation* is an instruction to move each atom of the molecule in a particular way with respect to a given element.

One important application of elements and operations is in determining whether a compound is optically active, that is, whether it will rotate the plane of plane-polarized light. Another use is in describing a molecule in terms of its stereochemical characteristics.

Reflection Operations

Element	Molecule with Element	Result of Reflection Operation
Point	 H ——•—— Cl	 Cl ——•—— H
Point	CH_3, CH_3, H, H	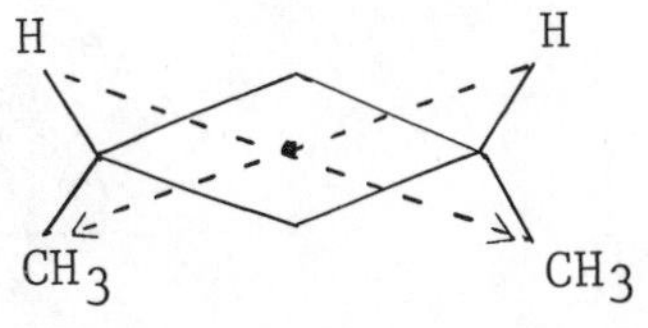 H, H, CH_3, CH_3
Plane	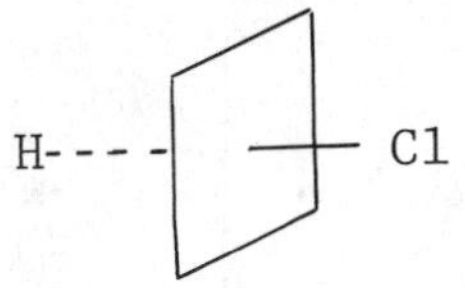 H- - - Cl	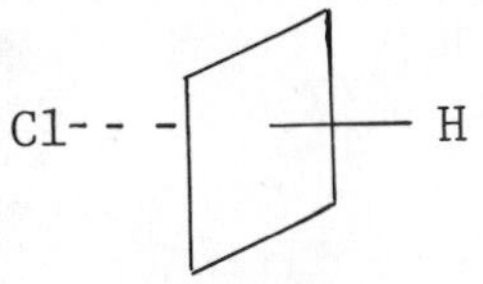 Cl- - H
Plane	CH_3, Cl, H, H	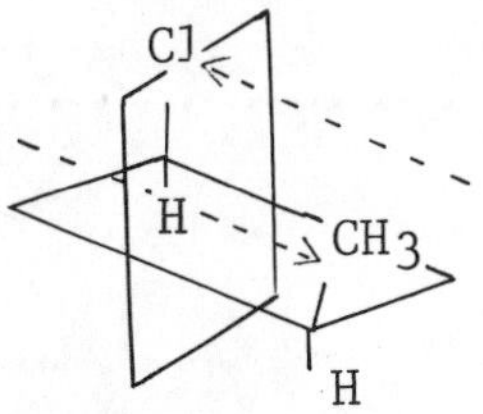 Cl, H, CH_3, H
Plane	CH_3, Cl, H, H	Cl, H, CH_3, H

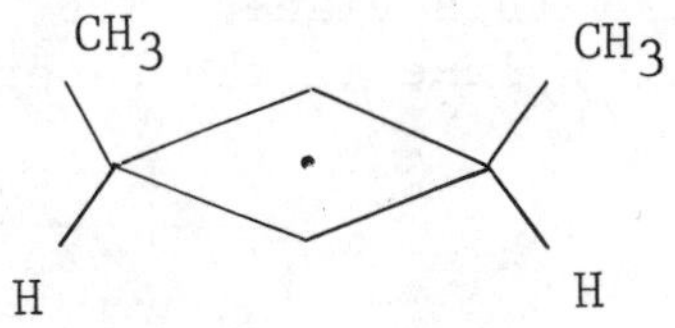

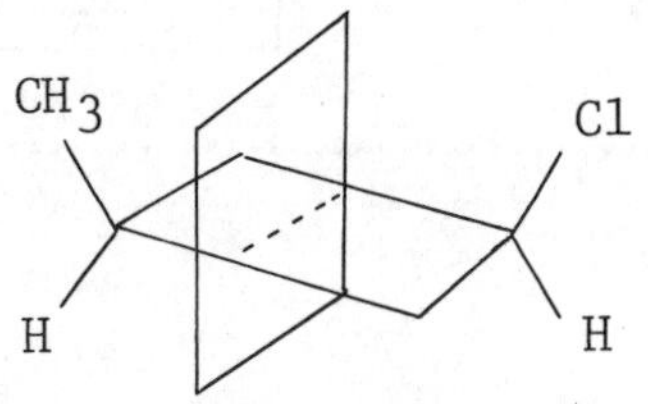

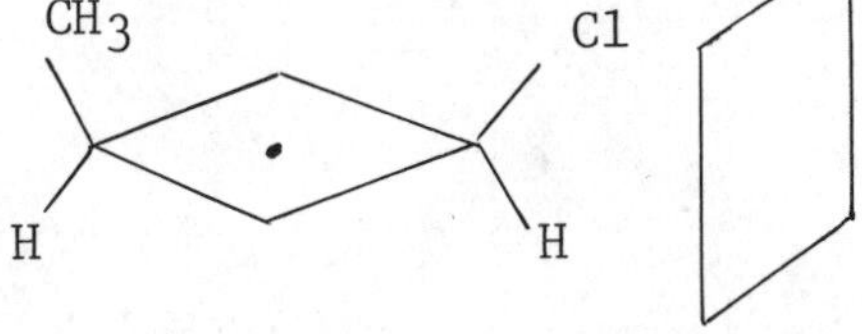

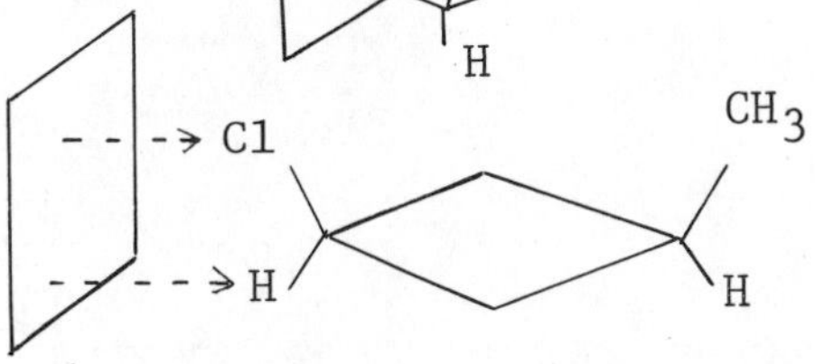

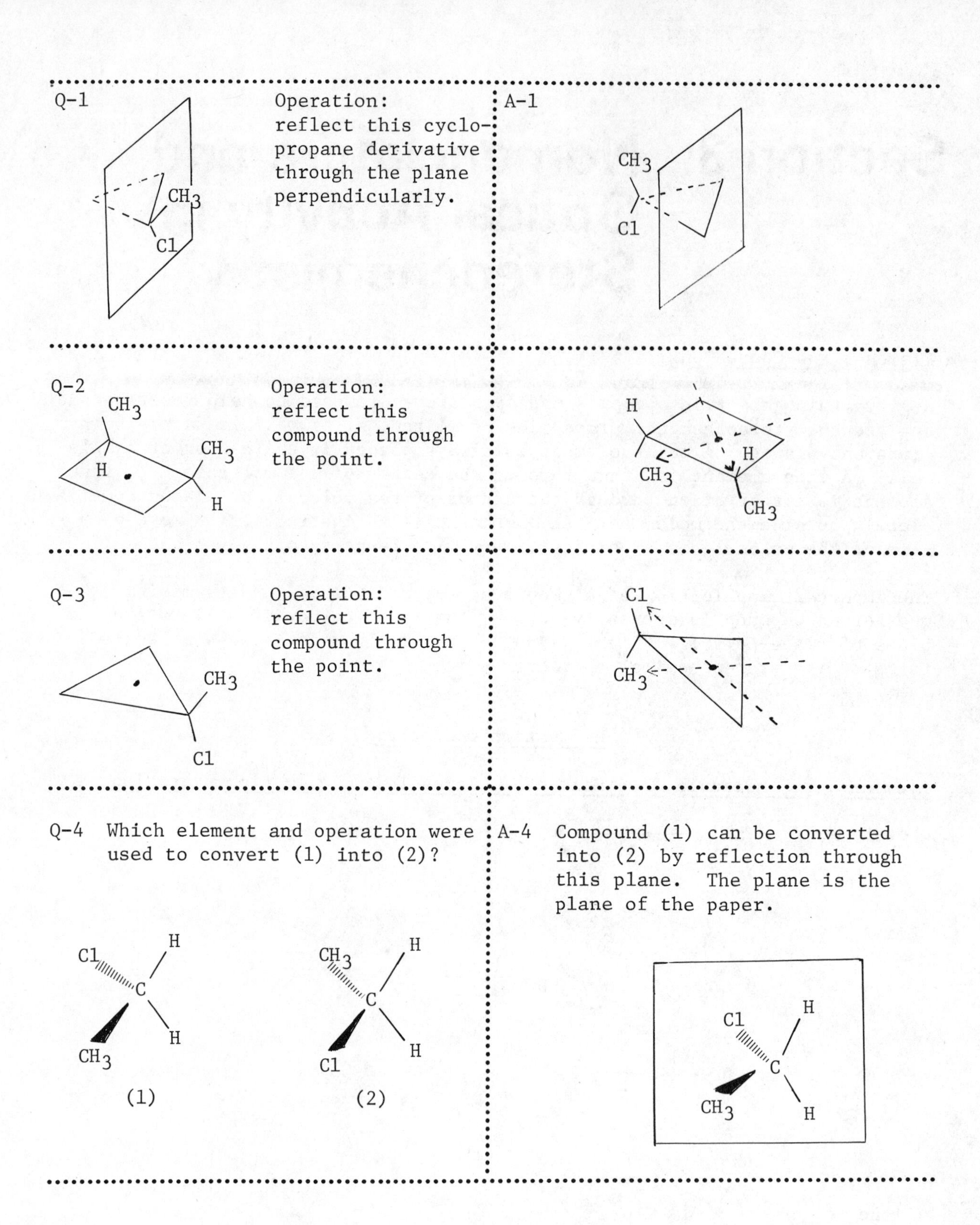
Q-1
CH3
Cl
Operation:
reflect this cyclo-
propane derivative
through the plane
perpendicularly.
A-1
CH3
Cl
Q-2
CH3
CH3
H
H
Operation:
reflect this
compound through
the point.
H
H
CH3
CH3
Q-3
CH3
Cl
Operation:
reflect this
compound through
the point.
Cl
CH3
Q-4 Which element and operation were
used to convert (1) into (2)?
Cl
H
C
H
CH3
(1)
CH3
H
C
H
Cl
(2)
A-4 Compound (1) can be converted
into (2) by reflection through
this plane. The plane is the
plane of the paper.
Cl
H
C
CH3
H

Q-5 When an operation with respect to some element results in a structure which looks the same as it did before the operation, then the operation is called a "symmetry operation." Which of the following two operations is a symmetry operation?

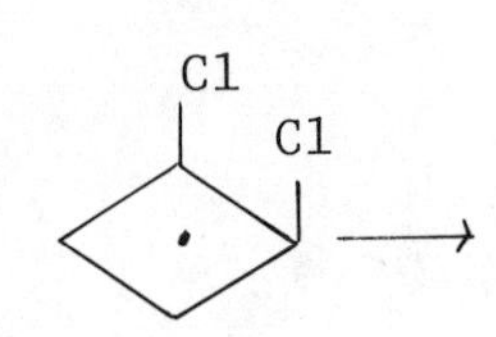

Operation A: reflect each atom through this point.

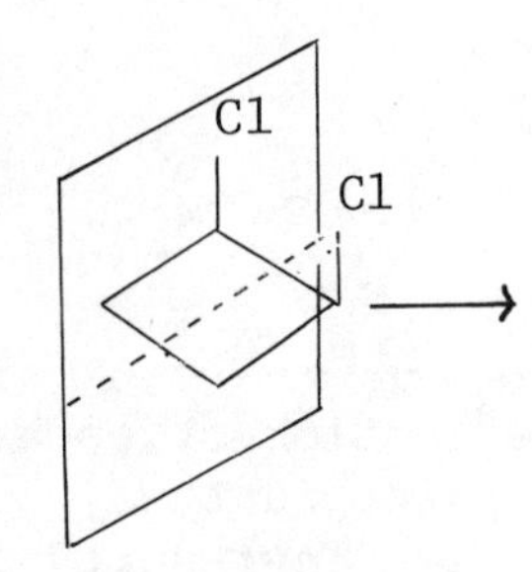

Operation B: reflect each atom through this plane perpendicularly.

A-5 Only operation B is a symmetry operation. Operation A results in a structure which looks different only because it has a different orientation.

Cl
Cl
Operation A
Cl
Cl

During a symmetry operation, each atom is either: a) moved to a position which was initially occupied by the same type of atom or group, or b) is not moved at all.

Q-6 The compound in Q-5 is said to have a "plane of symmetry." Explain. Does it also have a point of symmetry?

A-6 The compound has a plane of symmetry because reflection of each atom through this plane produces an identical structure. In other words, the operation with respect to this plane is a symmetry operation.

The compound does not have a point of symmetry because the operation with respect to this point does not result in an identical structure. This operation is thus not a symmetry operation.

An element is a symmetry element only when the corresponding operation is a symmetry operation.

Q-7 The symbol for a point (center) symmetry element is "i" and that for a plane symmetry element is "σ" (pronounced "sigma").

Label any symmetry elements shown in the following structures.

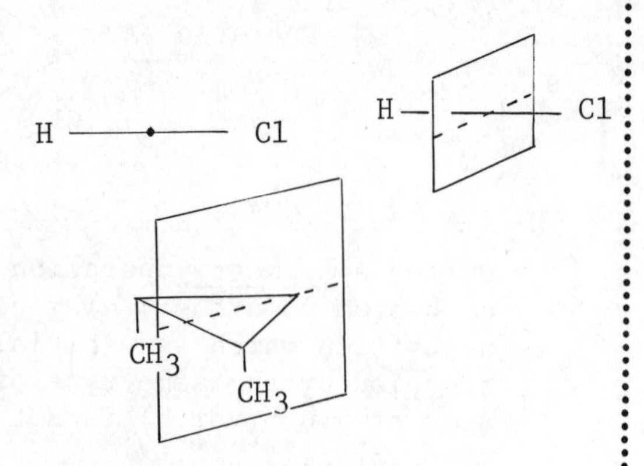

A-7 In order for a point to be labeled "i" or a plane to be labeled "σ" it must be a symmetry element. Only the plane for the cyclic compound is a symmetry element. It should be labeled σ .

Q-8 Does this compound have a center of symmetry? Explain.

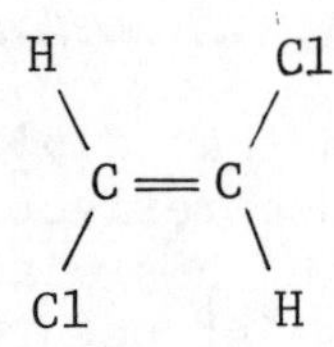

A-8 Yes, because reflection of each atom through a point in the center of the molecule results in a structure which looks the same.

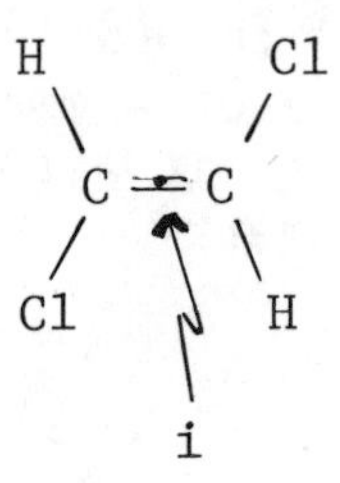

Q-9 Does this compound have a center of symmetry? Explain.

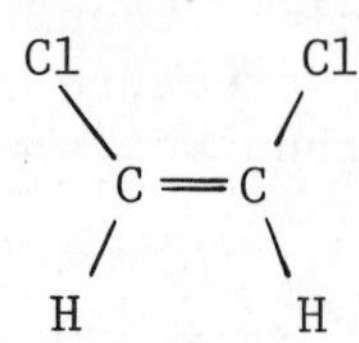

A-9 No, because reflection of each atom through a point in the center of the molecule results in a structure which looks different.

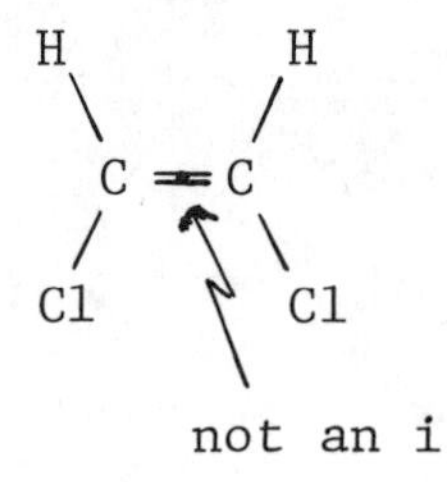

Q-10 Is this plane a plane of symmetry? Which of the atoms do not move when the molecule is reflected through this plane?

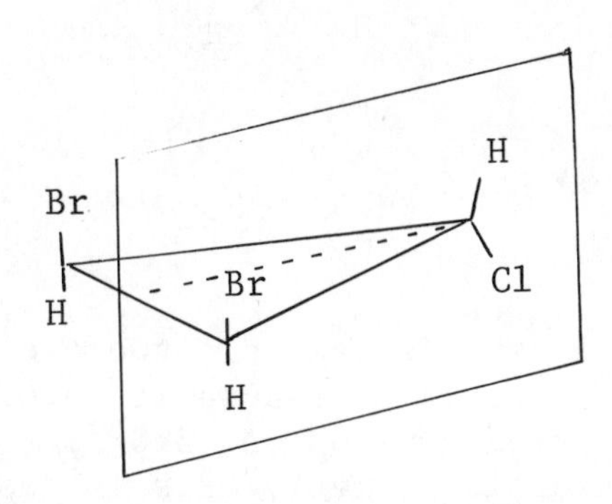

A-10 Yes, because reflection of each atom through this plane results in a structure which looks the same.

Only the three, joined H-C-Cl atoms do not move.

Q-11 Water has two elements of symmetry. Draw them.

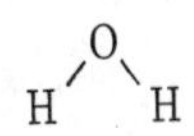

A-11

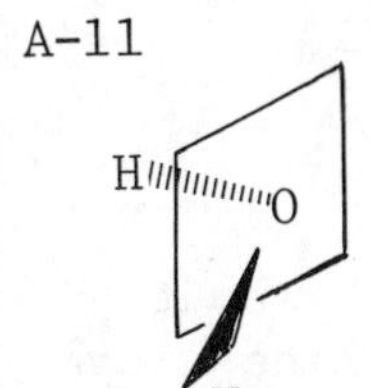

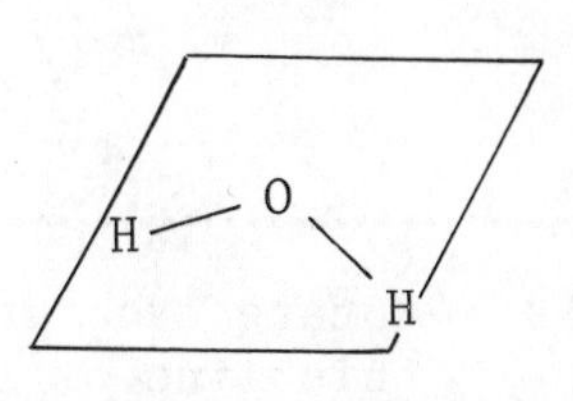

The two H's exchange position upon reflection.

All three atoms are in the plane. None of them move upon reflection.

Q-12 Lines (axes) are the third type of element used in conjunction with the operations. Carry out the following operation.

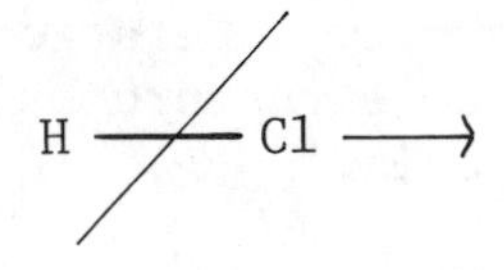

Operation: rotate each atom 180° about this line

A-12

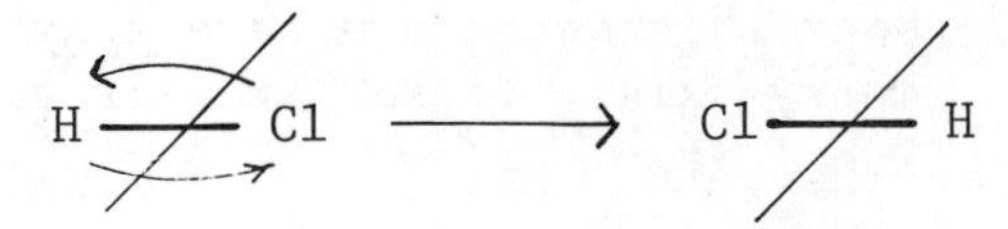

Q-13 Would you expect the line in Q-12 to constitute a symmetry element? Explain.

A-13 No, because the operation is not a symmetry operation. (The resulting structure looks different from the starting structure.)

Q-14 For which two angles of rotation in the range of 0° to 360° would the resultant structure in Q-12 look the same as the starting structure?

A-14 Rotations by 0° and 360° would do this. However, these rotations are not useful in stereochemistry.

Q-15 Is this a symmetry operation?

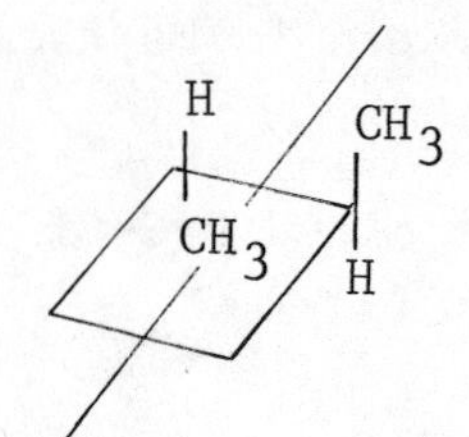

Operation: rotate each atom 180° about this line.

A-15 Yes, because the resultant structure looks the same as the starting structure.

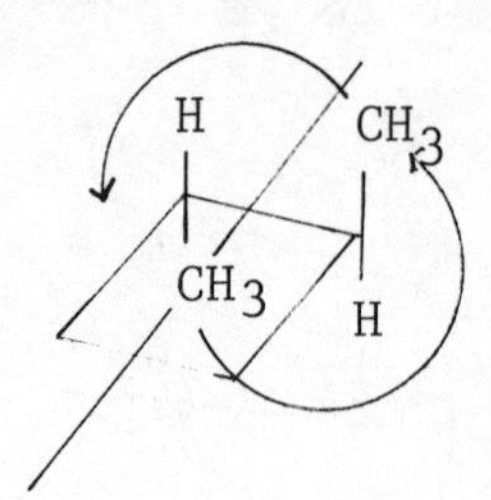

This axis is called a twofold symmetry axis because it involves rotation by 180°, and 360°/2 = 180°. A twofold symmetry axis is designated "C_2". An n-fold symmetry axis is "C_n" and called an n-fold rotation axis.

Q-16 Rotate each atom 180° about this line. Is this line a symmetry axis?

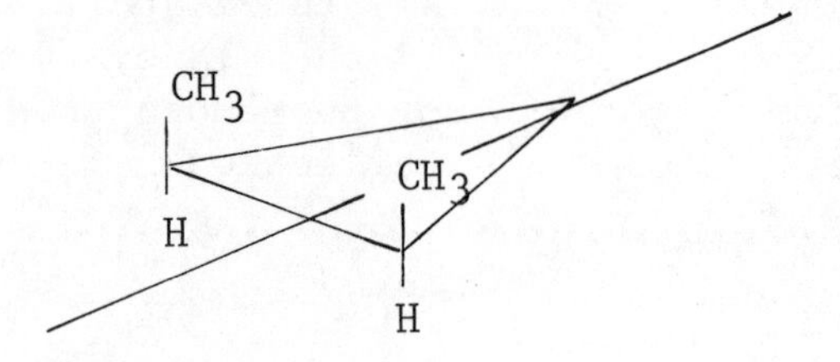

A-16 No, because some of the atoms have been moved to positions not initially occupied by the same type of atom. The resulting structure looks different from the initial structure.

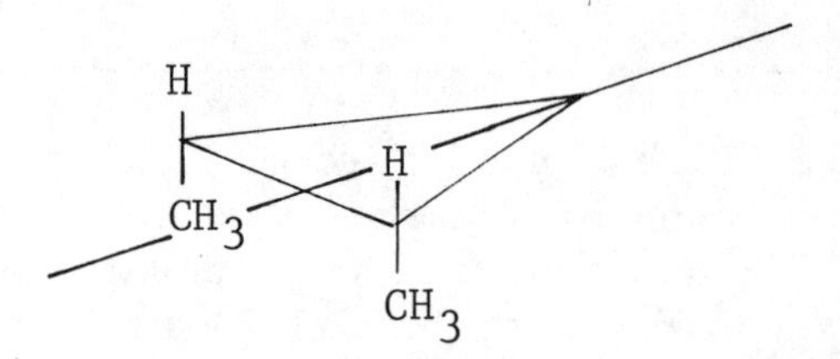

Q-17 Does this structure have a symmetry axis? If so, draw it and label it.

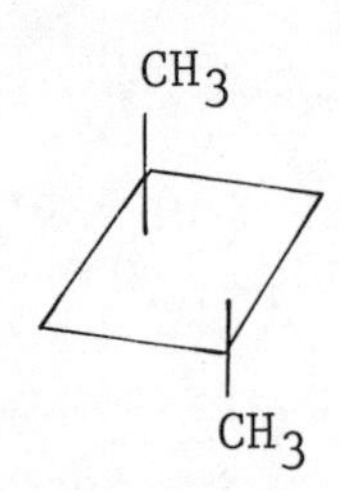

A-17 Yes, the symmetry operation is to rotate the structure by 180° about this axis.

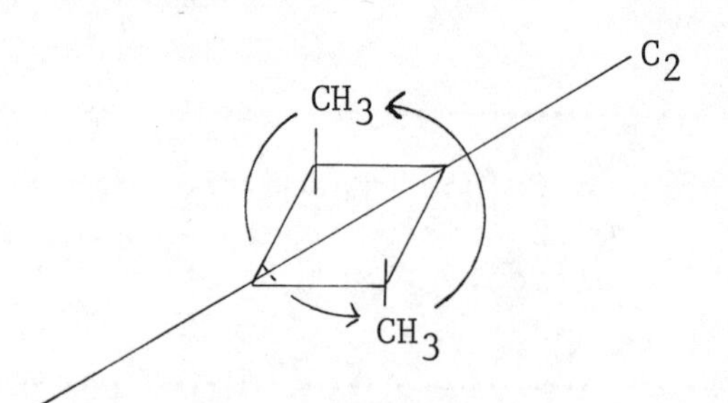

The axis is a twofold symmetry axis since 360°/2 = 180°.

Q-18 For which two angles of rotation (excluding 0° and 360°) is this line a symmetry axis? Give the appropriate C_n symbols for this axis.

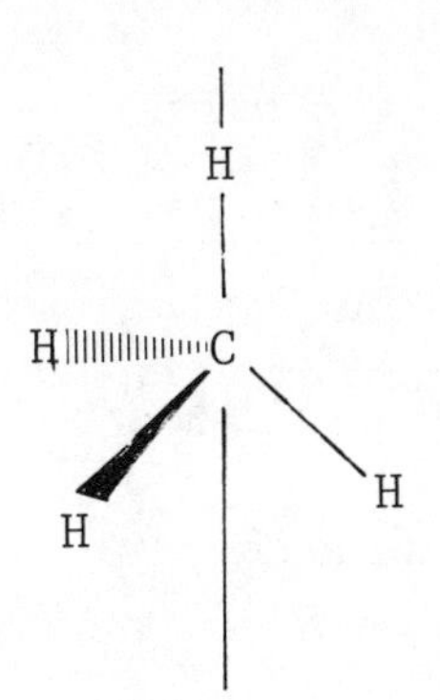

A-18 Rotations by 120° and 240° result in the initial structure. The 120° rotation is shown.

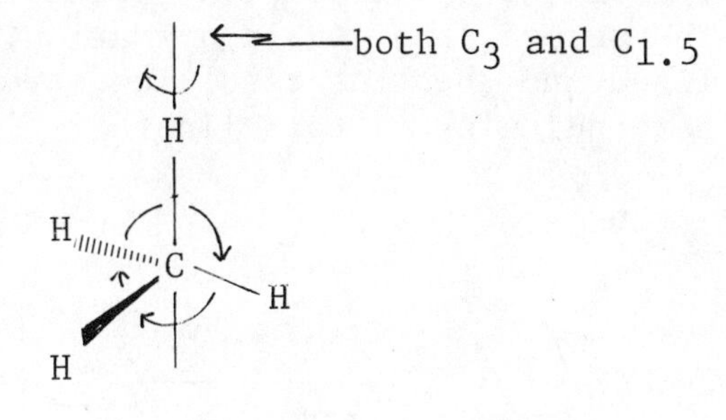

The line is both a threefold and 1.5 fold symmetry axis since $360^{o}/3 = 120^{o}$ and $360^{o}/1.5 = 240^{o}$.

Q-19 How many threefold axes does CH_4 have?

A-19 Four of them, one through each H.

Q-20 Consider a line which is perpendicular to the plane of the paper and passes through the central carbon atom in the drawing below.

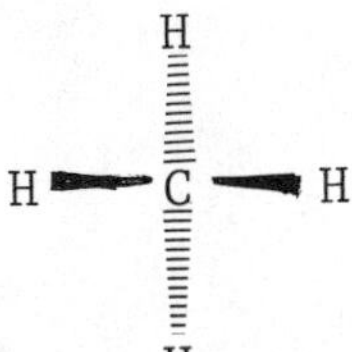

For which angles of rotation is the line a symmetry element?

A-20 Only rotation by 180° results in the initial structure. As a result, the axis is a two-fold axis.

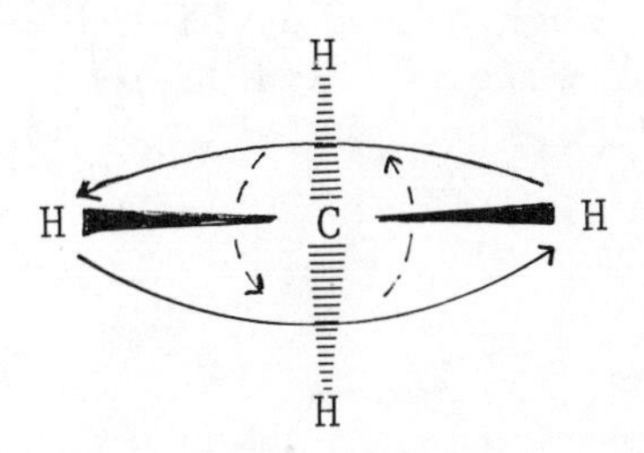

Q-21 How many of these twofold axes are there in CH_4?

A-21 Three. Each axis bisects two of the six sides of a tetrahedral representation of the molecule. The other two axes are shown below.

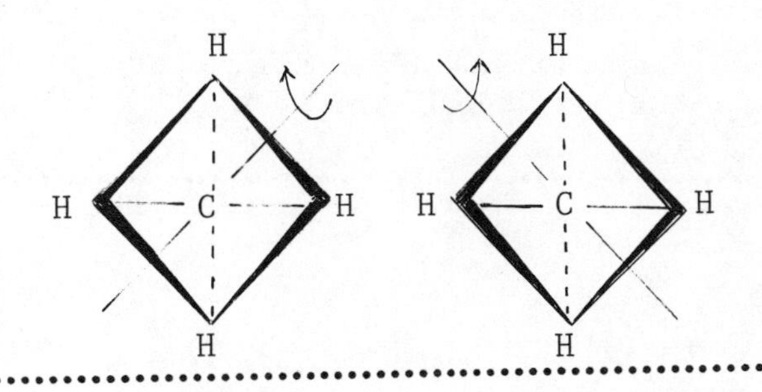

Lines as Symmetry Elements:

S-2 For lines as symmetry elements, two types of operations are performed. The first, rotation of a structure about a line, has already been presented. The second is a double operation. The structure is first rotated about a line, and then the resulting structure is reflected through a plane which is perpendicular to this line.

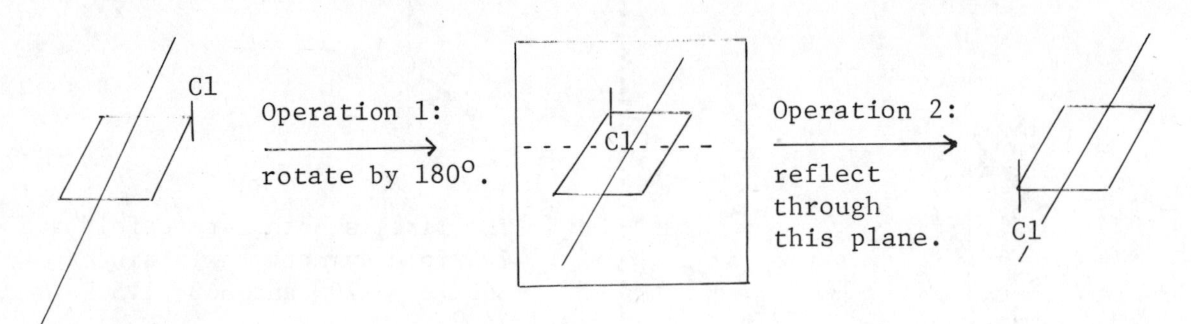

The line used in this symmetry operation is called an n-fold rotation-reflection axis and designated "S_n". The molecule is rotated 360/n degrees about this axis in the first operation.

Q-22 Rotate this structure 180° about the axis shown and then reflect the resulting structure through a plane which bisects the molecule and is perpendicular to the axis shown.

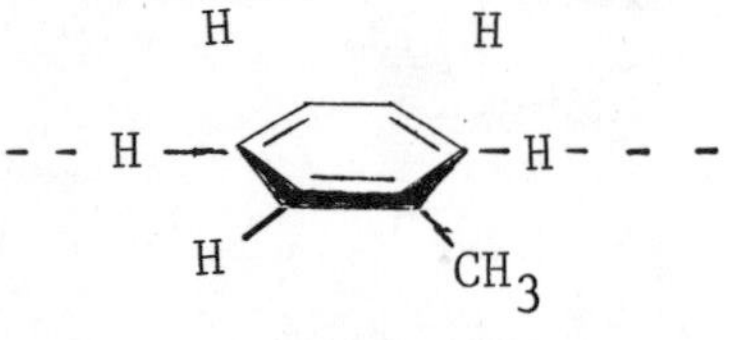

H H

-- H — — H -- -

H CH_3

(1)

A-22

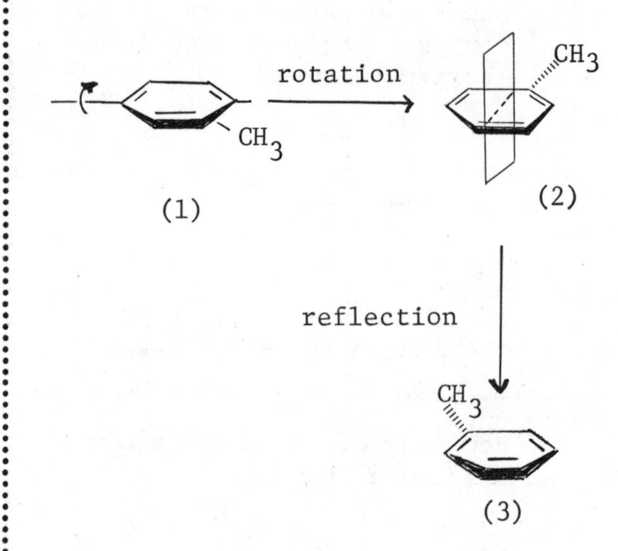

Q-23 Is the above operation (double operation) a symmetry operation? Explain.

A-23 No, because (3) is oriented differently from (1).

Q-24 Is the following operation a symmetry operation? If so, name and label the axis.

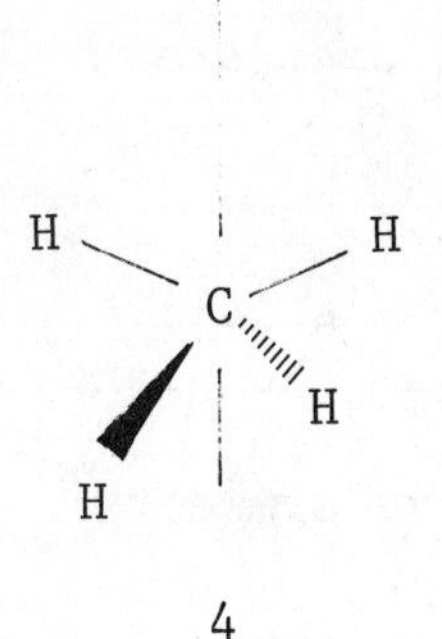

Operation: First rotate by 90° about the axis shown. Then reflect in a plane bisecting the molecule and perpendicular to the axis shown

A-24 Yes, because rotation by 90° and reflection through a plane, as shown, transforms (4) into itself.

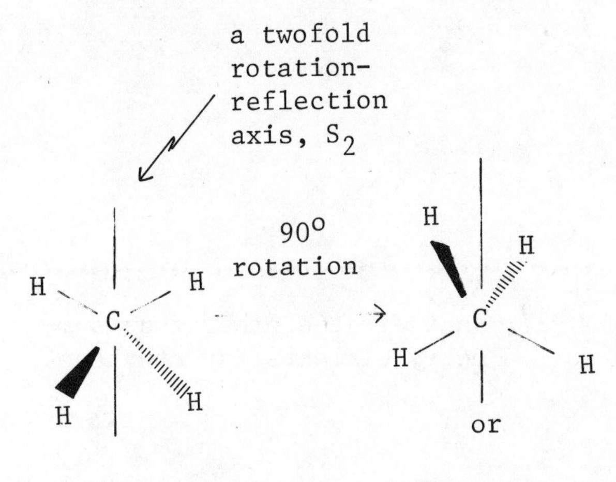

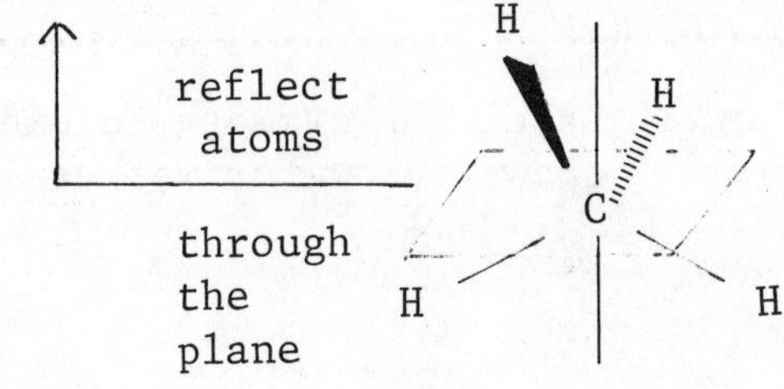

Q-25 What are the three symmetry elements for water? Give the symbols.

A-25 Water has two planes of symmetry (two σ) and one twofold axis (C_2). (C_2 means rotation by $360°/2 = 180°$ about the C_2 axis.)

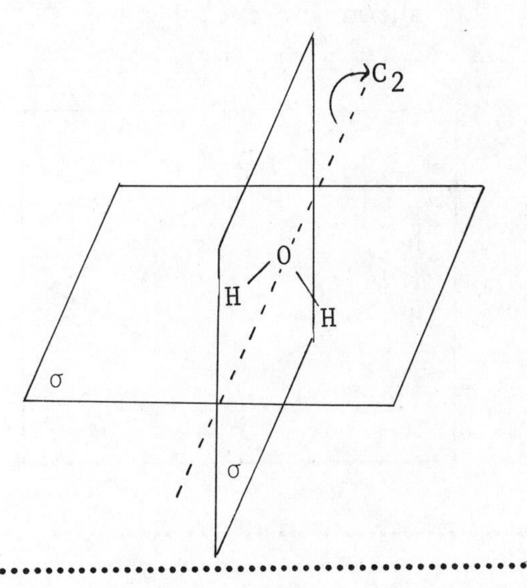

Q-26 Determine the three C_2 and three S_2 elements for ethylene.

A-26 C_2 and S_2

H C=C H ... C_2 and S_2

H H

C_2 and S_2

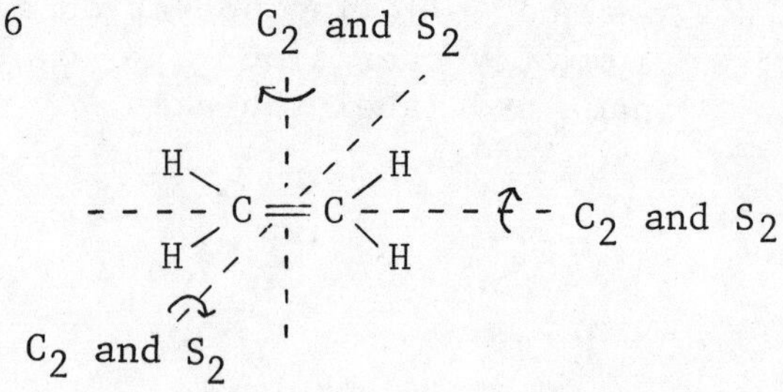

The three C_2 and S_2 axes are colinear.

Q-27 What are the other four symmetry elements for ethylene?

A-27 Ethylene has three σ's, all perpendicular (one is the plane of the paper) and an *i* in addition to the above symmetry elements.

Q-28 Label the symmetry element shown below for methane.

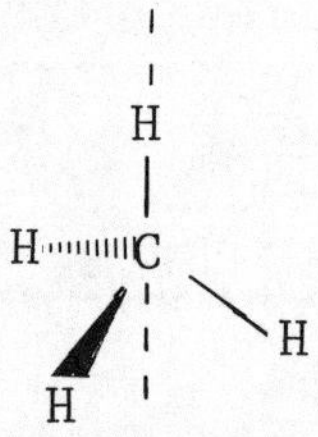

H
H C H
H

A-28 The vertical dotted line is a C_3 axis since rotation of methane by $360^o/3 = 120^o$ about this axis transforms the molecule into itself.

Q-29 How many C_3 axes does methane have?

A-29 Four of them, one through each hydrogen

Q-30 Label the two symmetry elements shown for cyclobutane.

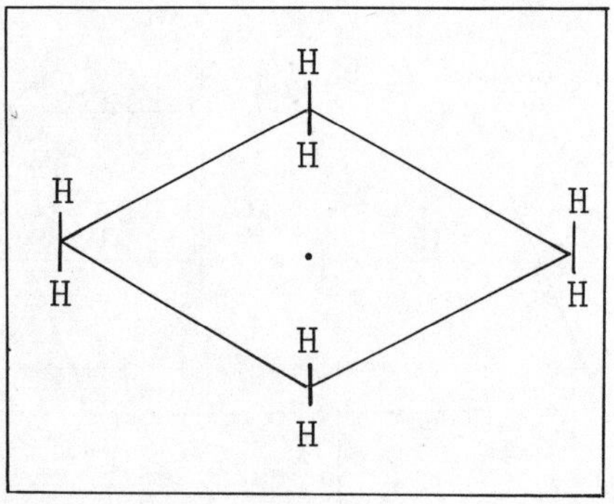

H
H
H H
H H
H
H

A-30

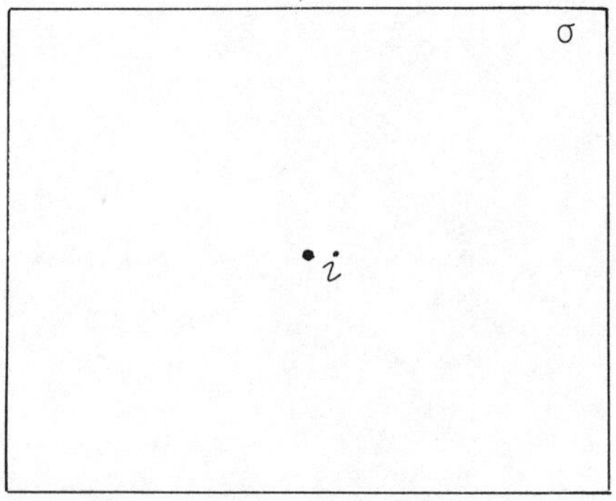

σ

•*i*

Q-31 How many planes does cyclobutane have?

A-31 Cyclobutane has five σ planes. Two pass through opposite corners (one of these is shown in Q-30), two pass midway through opposite sides, and one contains the four carbon atoms.

Q-32 Determine the four symmetry elements for ammonia (NH_3).

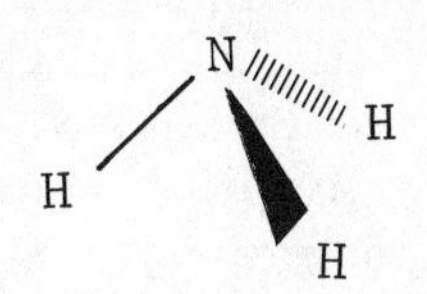

A-32

Ammonia has one C_3 axis (rotation by $360^o/3 = 120^o$) and three planes, each of which bisects an N-H bond and the C_3 axis. One of these is the plane of the paper.

Q-33 Determine the five symmetry elements for *trans*-1,2-dichloroethylene.

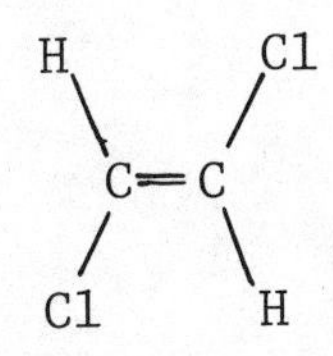

A-33

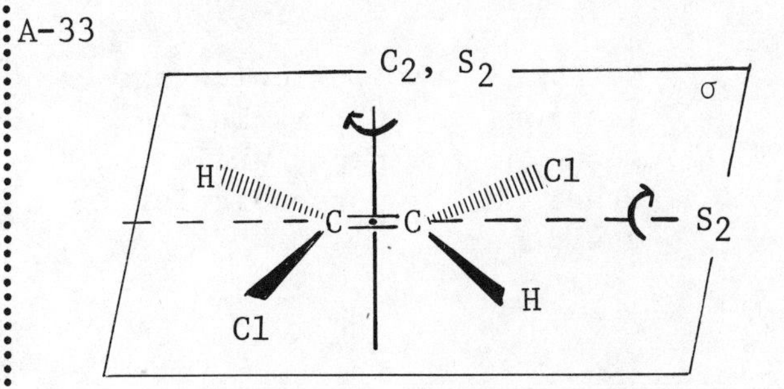

The compound has i, one C_2, one σ (plane), and two S_2 axes.

Q-34 1,2-dichloroethane, in the conformation shown below, has but a single symmetry element. What is this element. (Hint: draw a Newman projection.)

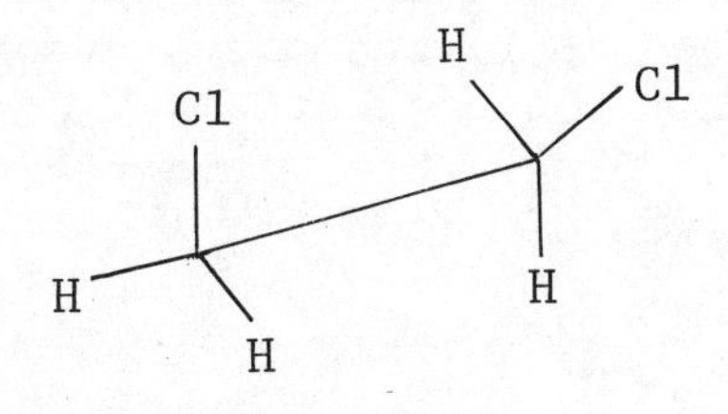

A-34

or

Q-35 What are the 24 symmetry elements of benzene?

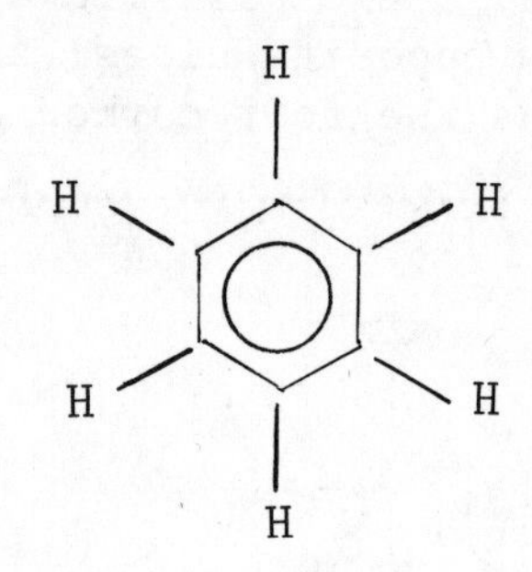

A-35

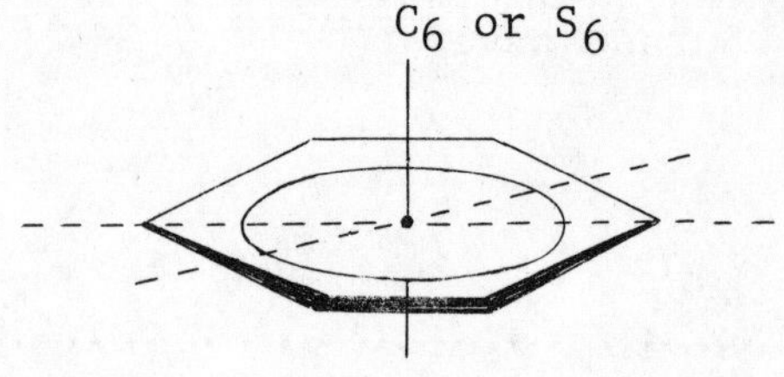

Benzene has one i, seven σ's. (three bisect H's, three pass in between adjacent H's, one is in the plane of the ring) one C_6, one S_6 (the C_6 and S_6 axes are colinear) six C_2's, six S_2's, one C_3 and one S_3.

Q-36 Allene has six symmetry elements. Three of these are C_2 axes, one of which is shown. Determine the other two C_2's. (Hint: draw a Newman-type projection.)

Note: In the following projection, one CH_2 is in the plane of the paper while the other CH_2 is perpendicular to the paper.

H H

— — C=C=C — — C_2

H H

A-36

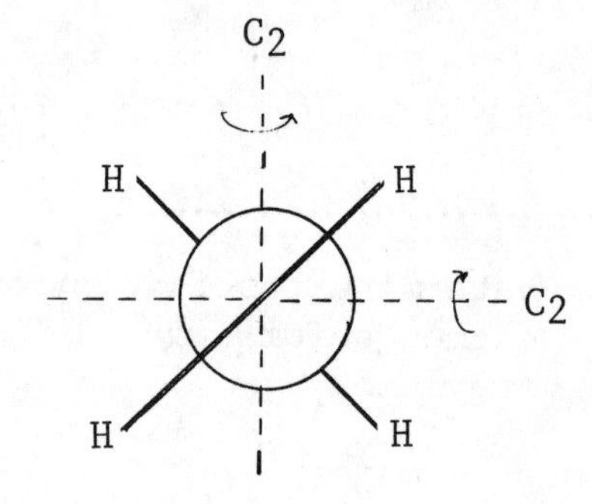

Q-37 Determine the other three symmetry elements of allene.

A-37

H H

— — C=C=C — — — S_4

H H

Two σ's (one is the plane of the paper) and one S_4 (colinear with one of the C_2's) comprise the other three symmetry elements of allene.

Asymmetry and Chirality:

S-3 The term *asymmetric* denotes absence of any symmetry. An object, such as a molecule in a given configuration, is termed asymmetric if it has no element of symmetry--that is, no i, σ, C_n or S_n.

The term *chiral* (kī́ -răl, "handedness") denotes absence of all symmetry, except at most axes of rotation (C_n), where $n > 1$. This term was adopted to define the structural features necessary for a substance to be optically active. All asymmetric molecules are chiral. Some symmetric molecules are also chiral since their symmetry is limited to rotation axes (C_n). In other words not all chiral molecules are asymmetric.**

Another term which has the same meaning as *chiral* is *dissymmetric*. Substances which are not chiral (dissymmetric) are called *achiral*.

Q-38 Is compound shown below asymmetric? Explain.

(1)

A-38 No, since all the atoms lie in the plane of the paper, this plane is a symmetry element.

Q-39 Any carbon which is tetrahedrally bonded to four different atoms or groups is both a chiral and an asymmetric carbon atom or center.

Which of the following compounds has a chiral (asymmetric) carbon?

(2) $C_2H_5—C(H)(H)—CH_3$

(3) $H—C(Cl)(C_2H_5)—CH_3$

A-39 Only compound (3) because it has four different groups on one of its carbons. The chiral (asymmetric) center is labeled with a star (*).

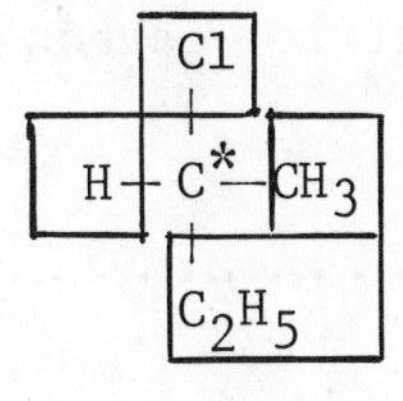

**See page 144.

Q-40 Does the compound shown in Q-38 have a chiral (asymmetric) carbon? How can you tell?

A-40 No, because none of the carbons is tetrahedrally bonded to four different groups.

Q-41 Does this compound have a chiral center? If so, label it with a star (*).

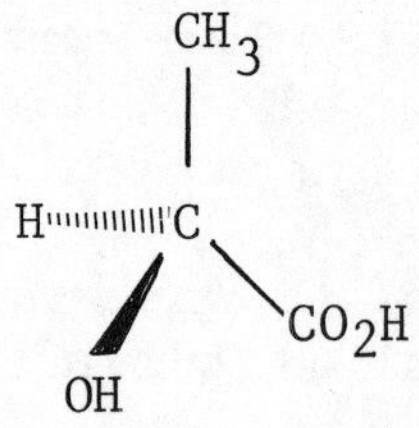

CH_3
H C CO_2H
OH

A-41 Yes, because four different groups are tetrahedrally bonded to one of the carbons.

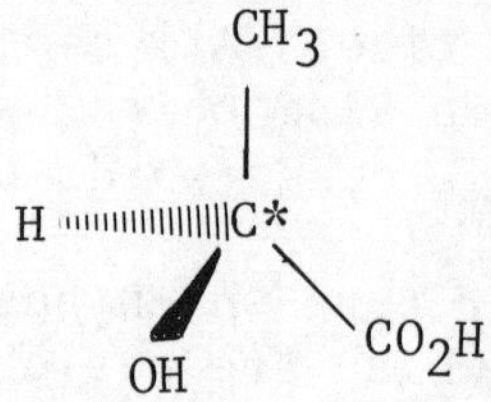

CH_3
H C*
OH CO_2H

Q-42 Does this compound have a chiral (asymmetric) center? Explain.

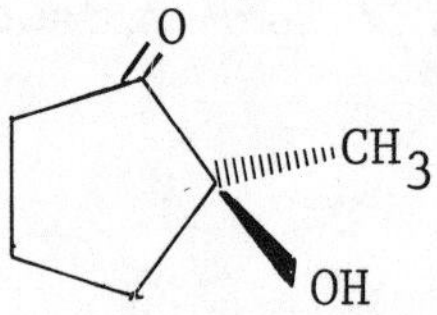

O
CH_3
OH

A-42 The starred carbon atom is chiral. The ring is considered to be two different groups since the order of substitution in one direction about the ring is different from that in the other direction.

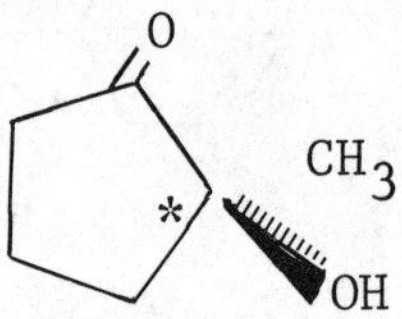

O
CH_3
*
OH

Q-43 Replace as many of the hydrogen atoms of methane with the halogen atoms F, Cl, Br as is necessary to make a chiral (asymmetric) carbon atom.

A-43

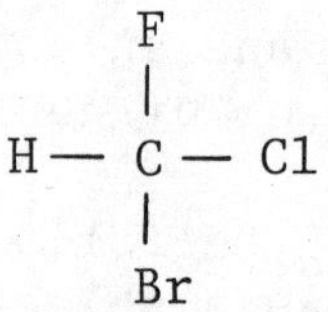

```
      F
      |
H — C — Cl
      |
      Br
```

Q-44 Star (*) the chiral (asymmetric) carbons in this compound.

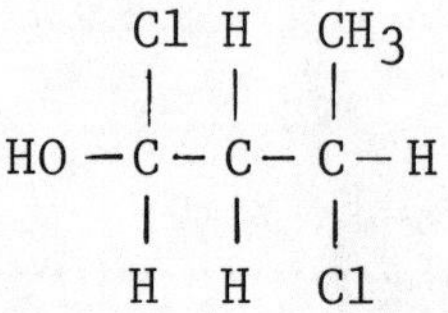

```
     Cl  H   CH3
     |   |   |
HO — C — C — C — H
     |   |   |
     H   H   Cl
```

A-44

```
     Cl  H   CH3
     |   |   |
HO — C* — C — C* — H
     |   |   |
     H   H   Cl
```

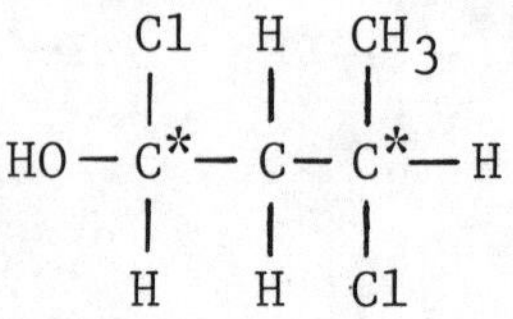

Q-45 Star the chiral (asymmetric) centers in this compound.

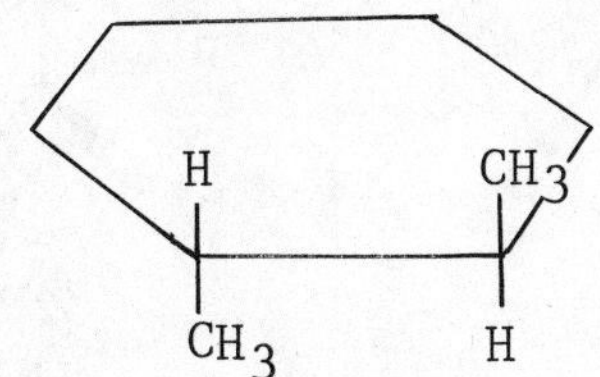

A-45

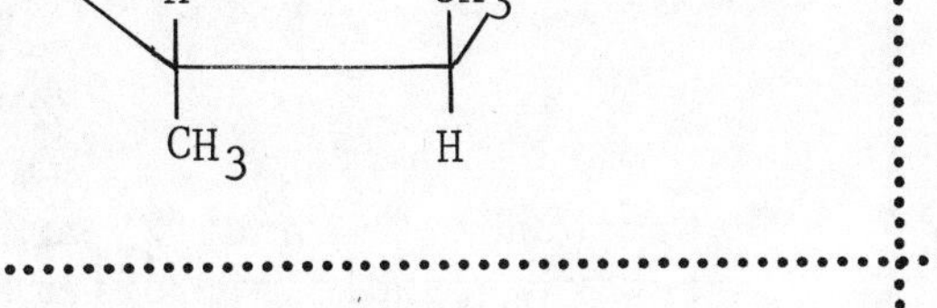

Q-46 Label the five chiral (asymmetric) carbon atoms in this compound.

A-46

Q-47 Label any chiral (asymmetric) carbon atoms in this molecule. Is the compound chiral? Asymmetric?

A-47 The compound has no chiral (asymmetric) carbon atoms (the order of substitution around the ring is the same in both directions) but it is both chiral and asymmetric because it possesses no symmetry elements.

Therefore, a compound does not have to possess chiral (asymmetric) carbon atoms to be chiral or asymmetric.

Q-48 Label any chiral (asymmetric) carbon atoms in this molecule. Is the compound chiral? Asymmetric?

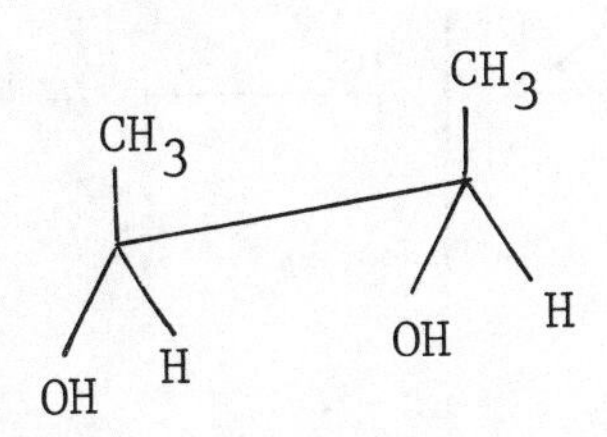

A-48 The compound has two chiral (asymmetric) carbons, but it is not chiral nor is it asymmetric because it has a plane of symmetry.

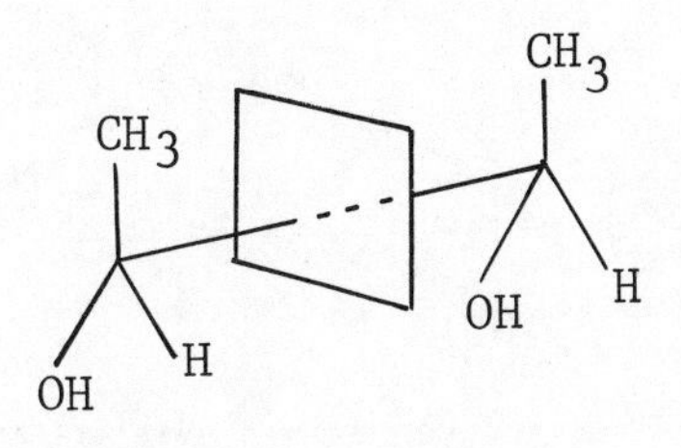

Such a compound is called a *meso* compound.

Q-49 A commonly used definition of chirality which is equivalent to that given in S-3 is the following: a chiral substance is not superimposable on its mirror image.

Draw the mirror image of compound (1) on the other side of the "mirror" and then determine whether (1) and its mirror image are superimposable.

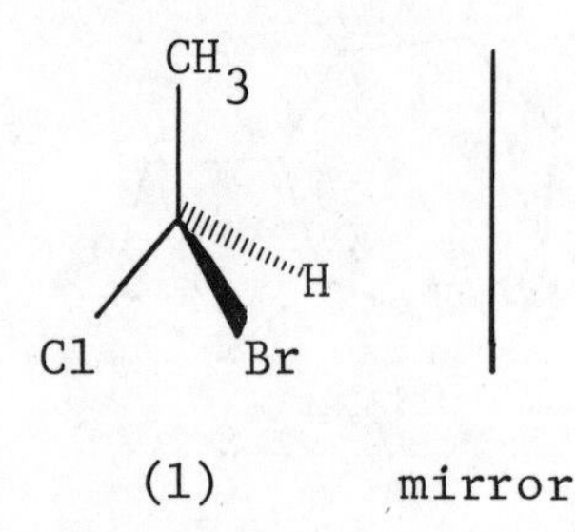

A-49

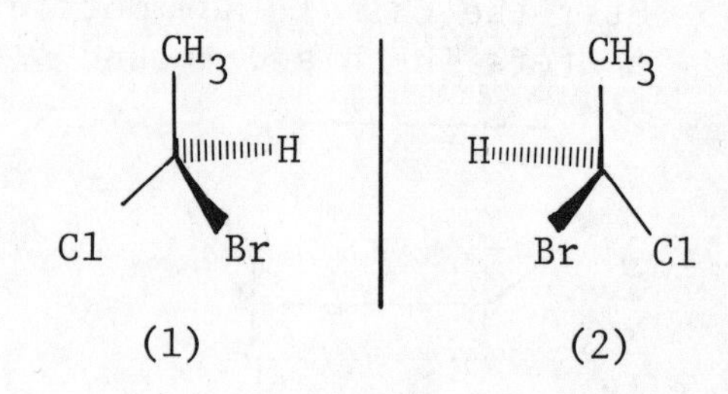

Compounds (1) and (2) are nonsuperimposable mirror images of each other. They are both chiral.

Q-50 Is this molecule chiral or achiral? Why?

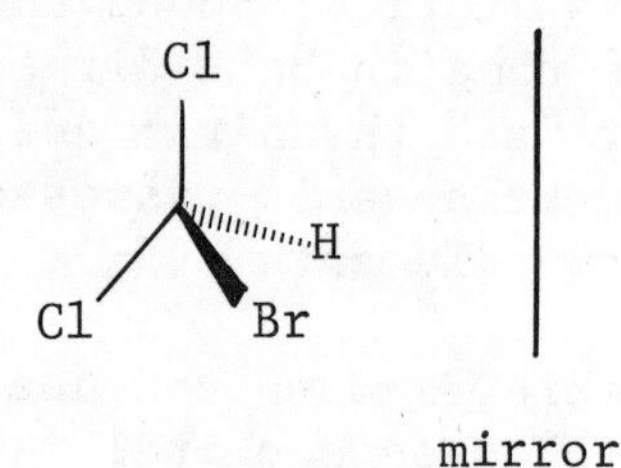

A-50 It is achiral for two reasons: (1) It has an element of symmetry other than a C_n element. (It has a plane of symmetry.) (2) It is superimposable on its mirror image.

Q-51 Is the molecule shown below chiral? Explain. Clues: (1) see if it is superimposable on its mirror image, (2) determine its symmetry elements, if any. Is the compound asymmetric?

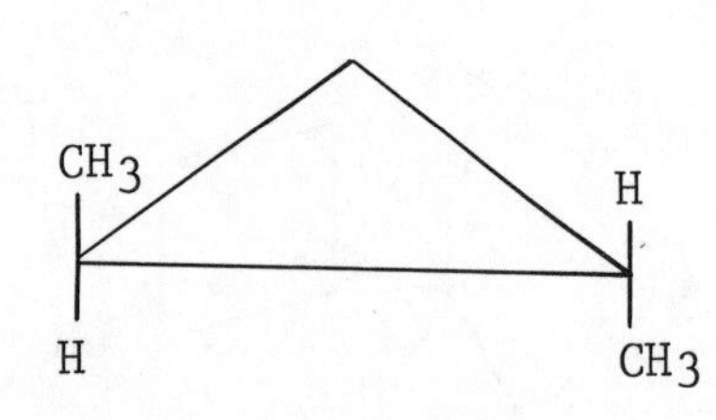

A-51 The molecule is chiral by both criteria: it is not superimposable on its mirror image and it possesses no symmetry elements other than a C_2 axis. But since it possess a C_2 axis, it is symmetric, not asymmetric.

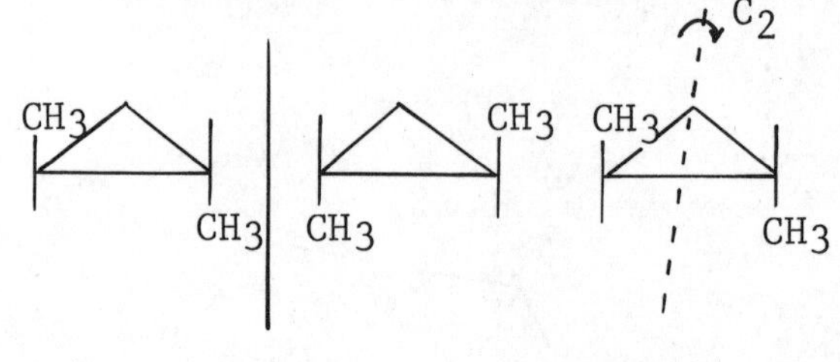

Q-52 Label the chiral centers in this compound

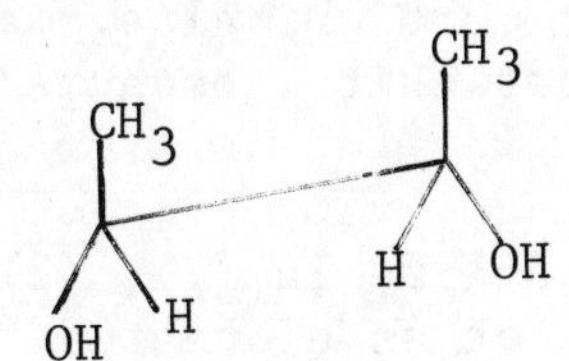

Is the compound chiral? Explain. Is it asymmetric? To evaluate how symmetrical a molecule is, that is, what symmetry elements a molecule possesses, consider the molecule in its most symmetric conformations.

A-52

The compound is chiral because it has no symmetry elements other than a C_2 axis. Because it has a C_2 axis, it is not asymmetric.

Q-53 Explain why the compound shown below is both symmetrical and achiral and yet possesses two chiral centers (*).

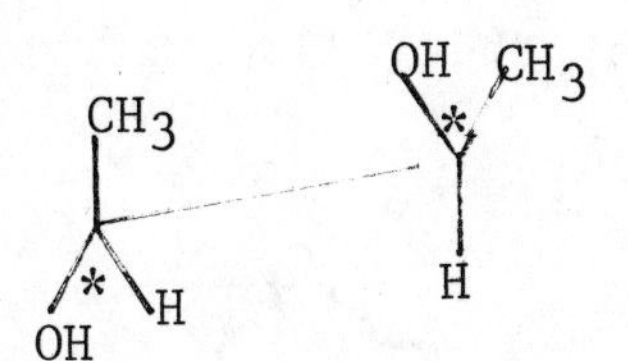

A-53 The compound is symmetrical and achiral for the same reason: it has one or more symmetry elements other than C_n axes (it has a plane of symmetry).

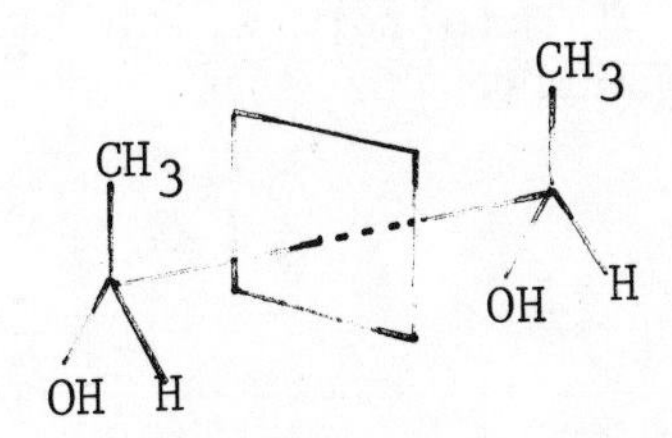

Chiral Elements:

S-4 The three elements of chirality are chiral centers, chiral axes, and chiral planes. All chiral compounds have one or more chiral elements, but some compounds possess chiral elements and are not chiral because they possess symmetry element(s) other than C_n.

A chiral axis is illustrated by the dotted line in (1), (2), and (3). In each drawing the dotted line is the long axis of an elongated tetrahedron. Note that there must be at least two different groups in corner positions and that the two groups at each end must be different.

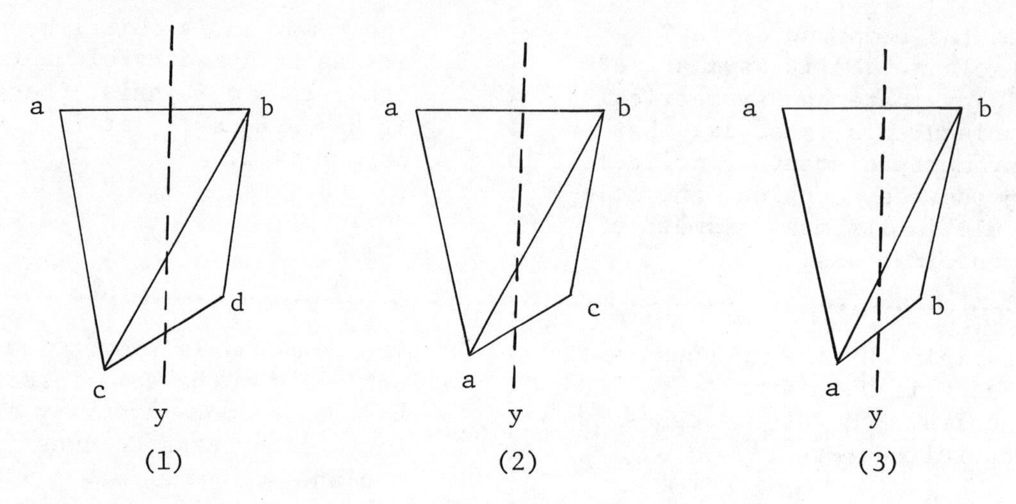

This form of chirality is exemplified by the following types of compounds.

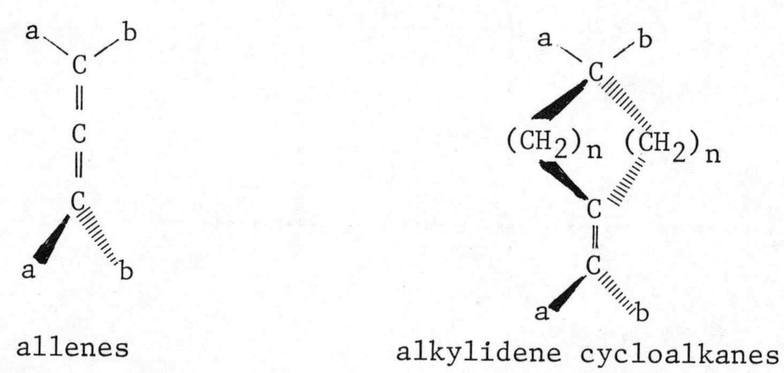

A chiral plane *z* is illustrated by the following structure in which some of the atoms of the molecule rigidly define a plane while the rest of the atoms form a loop above this plane. The structure is not superimposable on its mirror image.

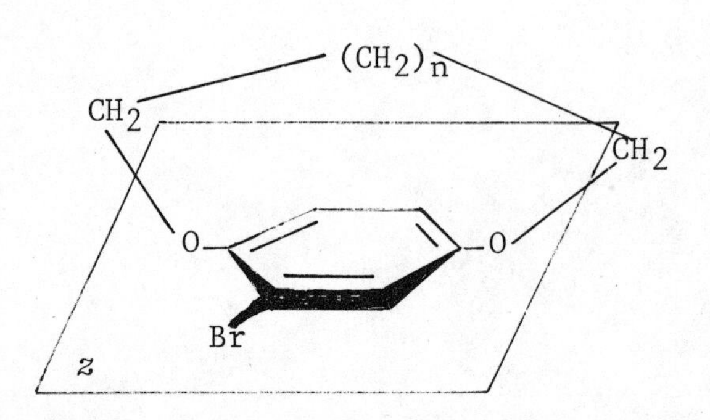

Q-54 Distinguish chiral elements from symmetry elements. (Both can be points, lines, and planes.)

A-54 A symmetry element is used to indicate the presence of symmetry with respect to the element.

A chiral element is used to indicate the absence of symmetry with respect to the element.

Q-55 Is the dashed line a chiral axis? Explain.

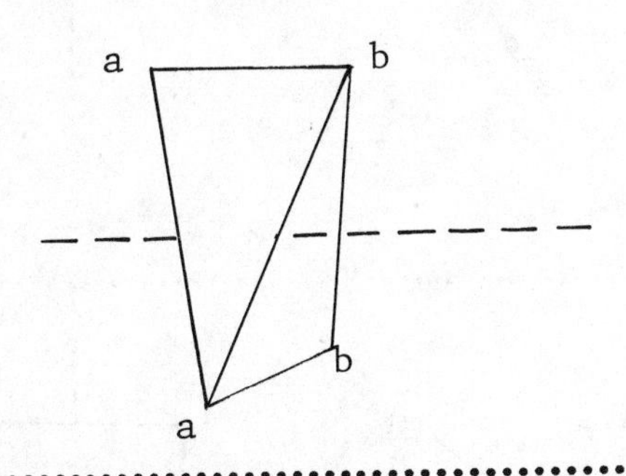

A-55 No, because it is not the long axis of the tetrahedron.

Q-56 Is this dashed line a chiral axis? Explain.

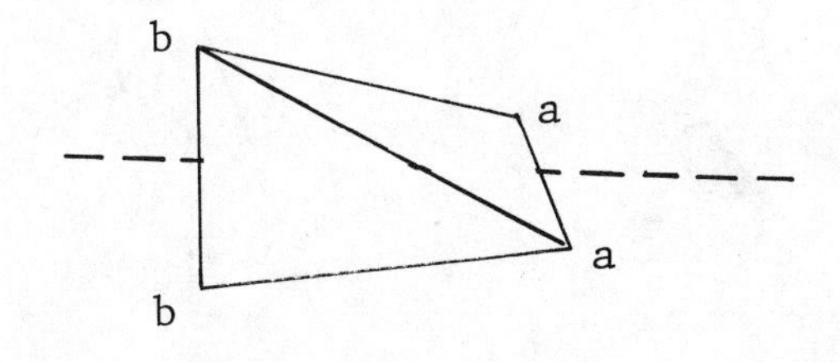

A-56 No, because at each end the two groups are the same.

Q-57 Does the compound shown below have a chiral axis? If so, draw it and label the type of substitution pattern.

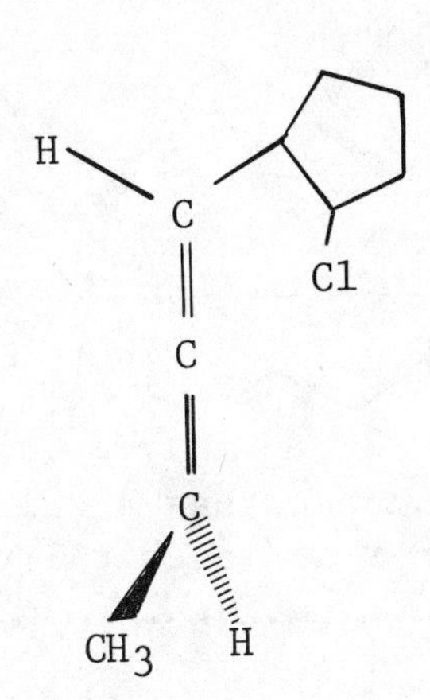

A-57 Yes, the chiral axis and substitution pattern are shown below.

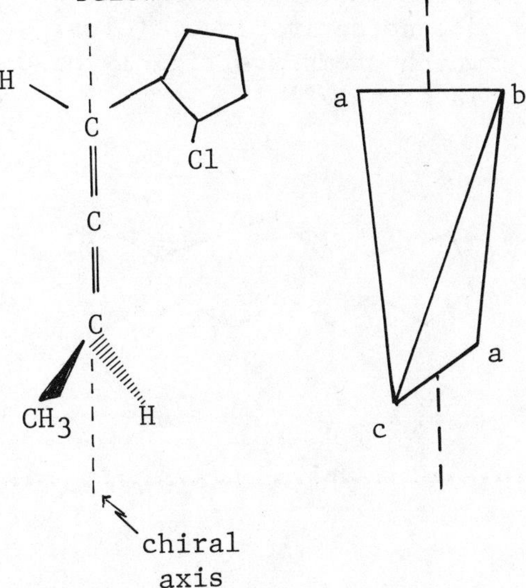

Q-58 Does the spiran shown below have a chiral axis? (Hint: redraw the compound to show three dimensional structure.)

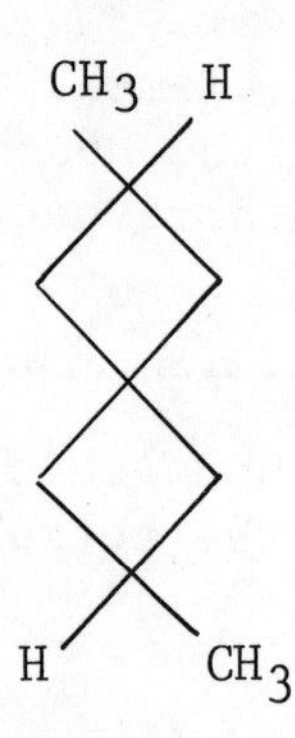

If so, what is the pattern of substitution?

A-58 Yes, there is a chiral axis as indicated below.

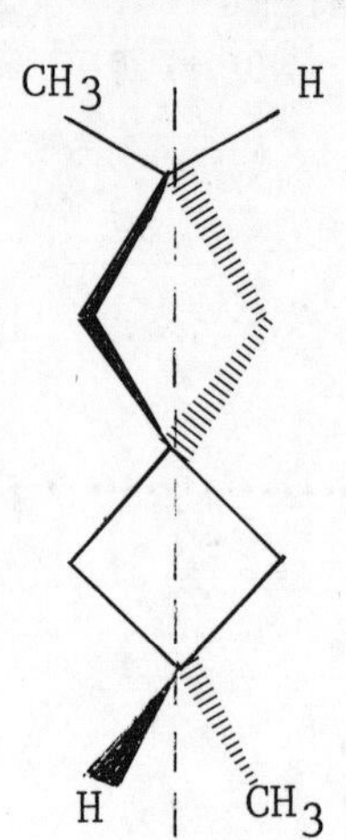

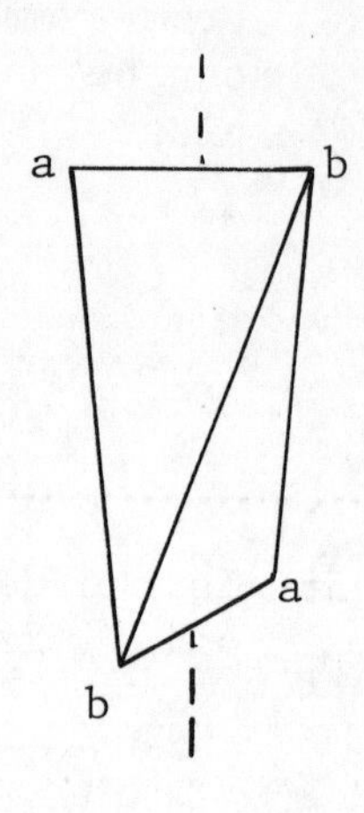

Q-59 Using CH_3 and Cl groups <u>place</u> groups on the structure shown below so that it acquires a chiral axis.

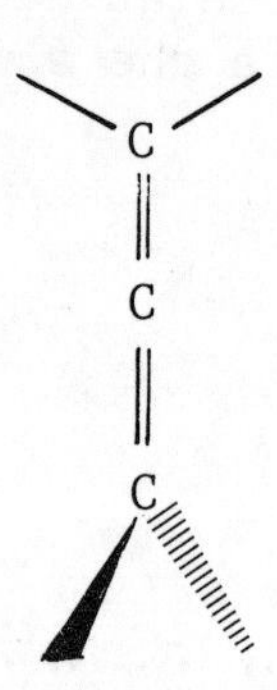

A-59

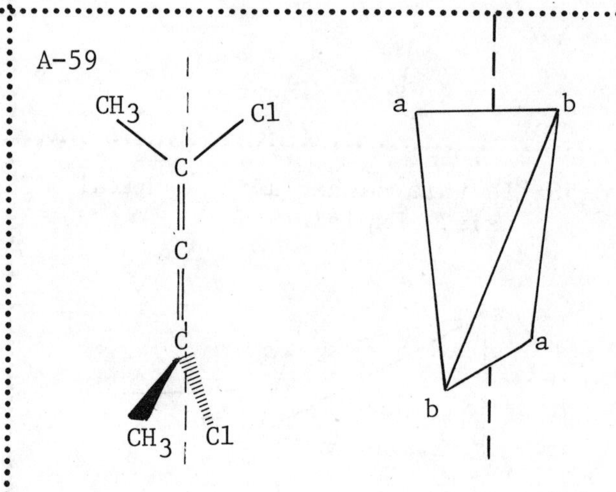

Q-60 Identify the two chiral elements in this *trans*-cyclododecene derivative. (Clue: which atoms are rigidly held in a plane?)

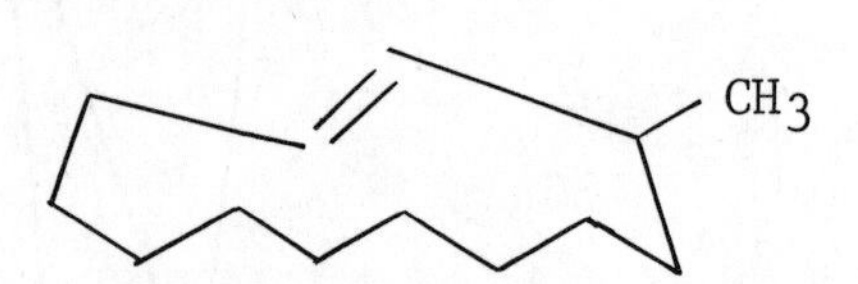

A-60

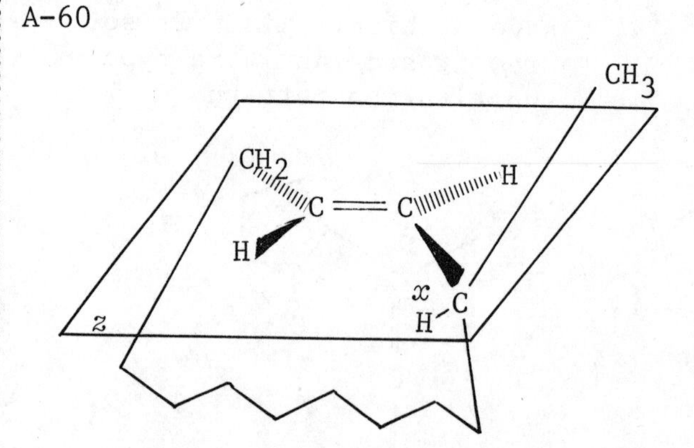

The compound has a chiral center *x* and a chiral plane *z*.

Q-61 Substitute four groups (a,a,b,b) on this polychlorobiphenyl conformer such that it acquires a chiral element. Draw the resulting chiral axis.

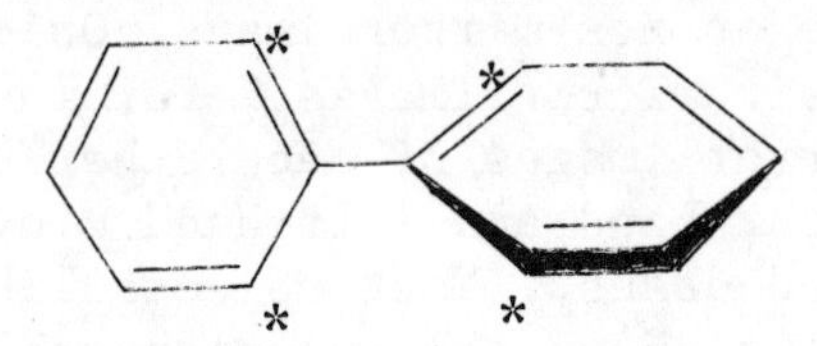

A-61

a
b
y
b
a

Q-62 Explain why this compound is both symmetrical and chiral.

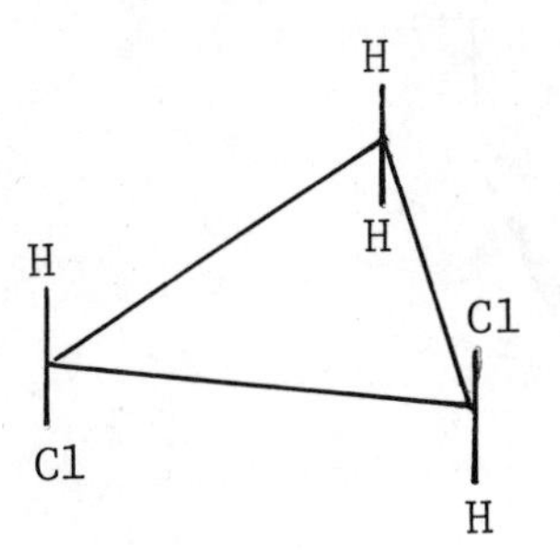

A-62 The compound is symmetric because it has a C_2 axis. The compound is chiral because it lacks all symmetry elements except, at most, a C_n, which is the C_2 axis.

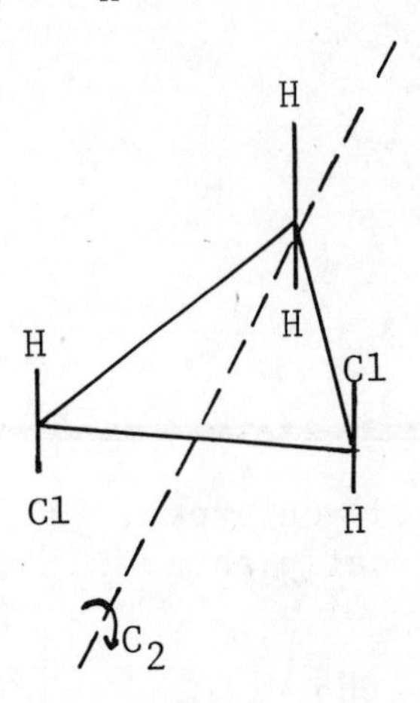

It is also chiral because it is not superimposable on its mirror image.

Optical Activity:

S-5 The most dramatic physical property of chiral substances is the capacity to rotate the plane of plane-polarized light. This process is termed "optical activity," and arises from the interaction of the plane-polarized light and the electrons of the chiral substance. For a molecule to be optically active it must be chiral, that is, not superimposable on its mirror image. Objects as well as molecules can be different from their mirror images. For instance, a person's hands (if perfectly shaped) are mirror images of each other but are not superimposable. The same goes for right-hand and left-hand gloves, right and left shoes, and right- and left-sided desks. Most objects like spoons, footballs and pullover sweaters, for instance, are superimposable on their mirror images (aside from any insignia, monograms, etc.).

Compound (1) is not superimposable on its mirror image (2) and is therefore optically active.

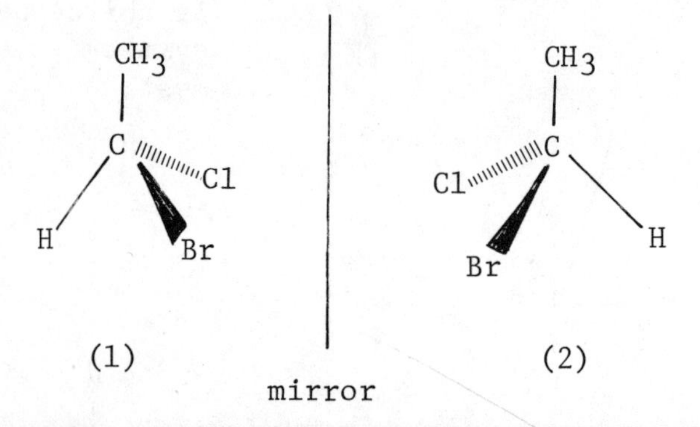

Q-63 Would you expect (3) to be optically active? Explain.

A-63 No, because it is superimposable on its mirror image.

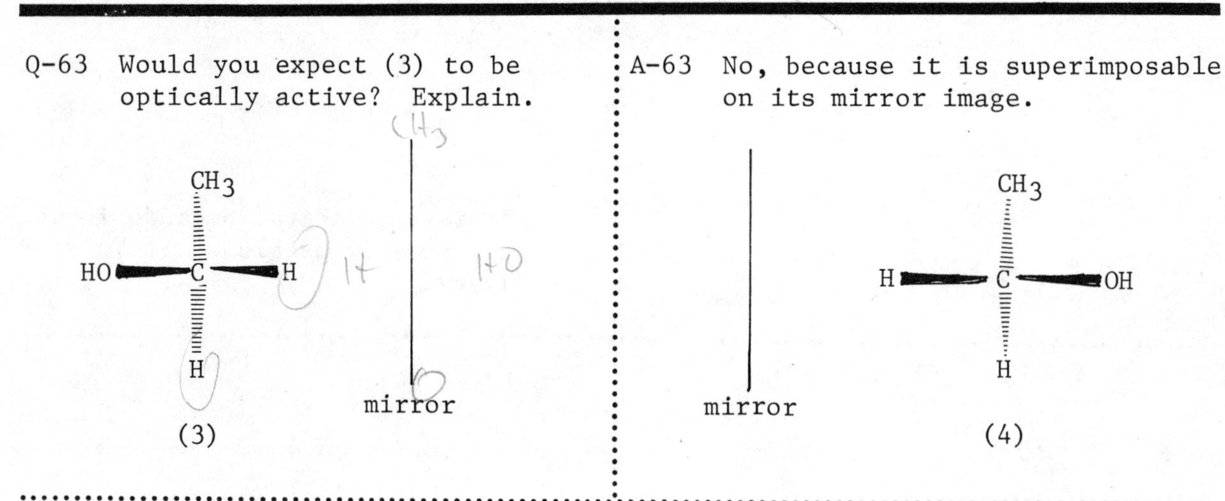

Q-64 Would you expect the compound (5) to be optically active? What about its mirror image?

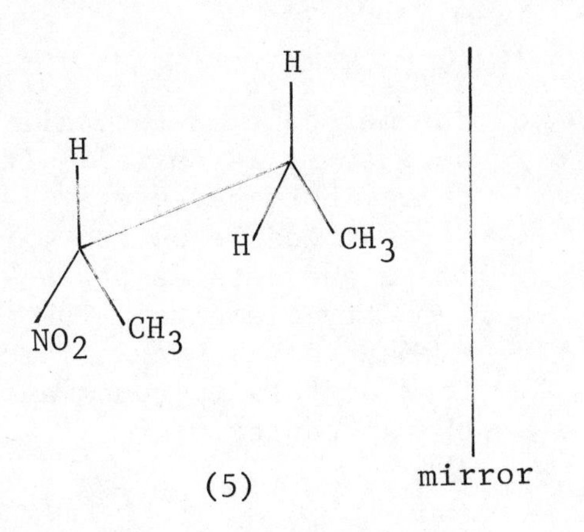

A-64 Both compounds are optically active because neither is superimposable on its mirror image. The compounds are mirror images of each other.

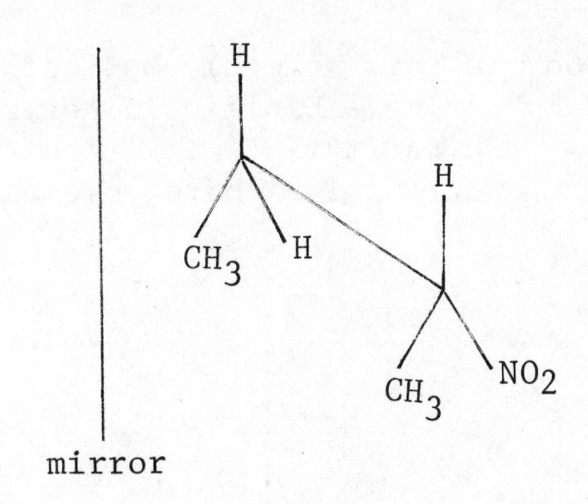

Optical Activity and Symmetry:

S-6 Although it is usually easy to draw the mirror image of a structure, it is sometimes a tricky exercise to test whether or not the mirror image is superimposable on the original. For this reason, other criteria, based on internal symmetry properties, are often used to predict the presence or absence of optical activity. The two symmetry criteria generally employed, as noted previously, are chirality and S_n symmetry.

Chirality expresses the necessary and sufficient condition for the existence of optical activity with a couple of unusual exceptions.* This means that any chiral molecule is optically active and that any achiral molecule is optically inactive.

The other generalized symmetry requirement for optical activity is the absence of an S_n axis ($n \neq 1$). Conversely, a molecule which possesses an S_n axis is necessarily optically inactive. The S_n criterion is easy to apply to molecules when n = 1 or 2 since S_1 is equivalent to a σ element (plane of symmetry) and S_2 is equivalent to an i element (center of symmetry).

When applying symmetry criteria, consider the most symmetric conformation of the molecule.

* A few molecules are not optically active even though they are chiral and lack an S_n axis (the two symmetry criteria for optical activity). The reasons are oblique: either the optical activity of the molecules is apparently too weak to be measured by currently available instrumentation, or the molecules may be considered to be unresolvable racemic mixtures. (Carlos, J. L., Jr., *J. Chem. Educ.* 45, 248 [1968].)

Q-65 Are all asymmetric molecules optically active? Explain.

A-65 Yes, because asymmetric molecules lack all symmetry elements. All asymmetric molecules are chiral.

Q-66 Is this molecule optically active? Explain in terms of chirality. If it is chiral, what is its chiral element?

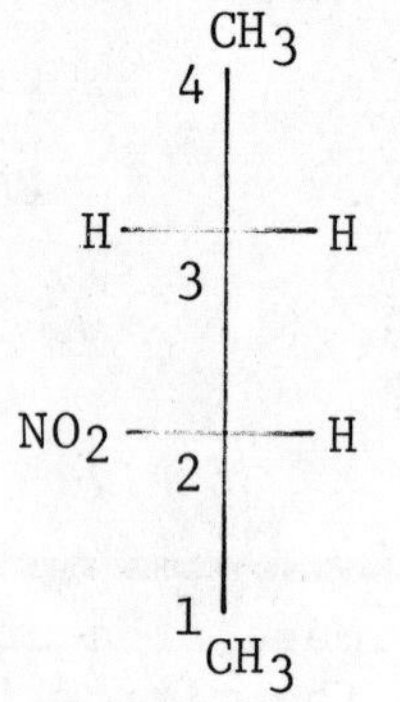

$$\begin{array}{c} CH_3 \ (4) \\ | \\ H - C_3 - H \\ | \\ NO_2 - C_2 - H \\ | \\ CH_3 \ (1) \end{array}$$

A-66 The molecule is optically active because it is chiral. It is chiral both because it is not superimposable on its mirror image and because it lacks all symmetry elements. The chiral element is carbon-2, which bonds four different groups and is a chiral center.

Q-67 Is this compound optically active? Explain in terms of S_n criteria.

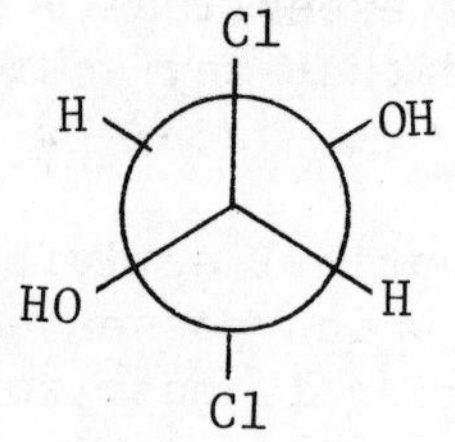

A-67 No, it is optically inactive because it possesses an S_1 axis (plane of symmetry) and an S_2 axis (center of symmetry).

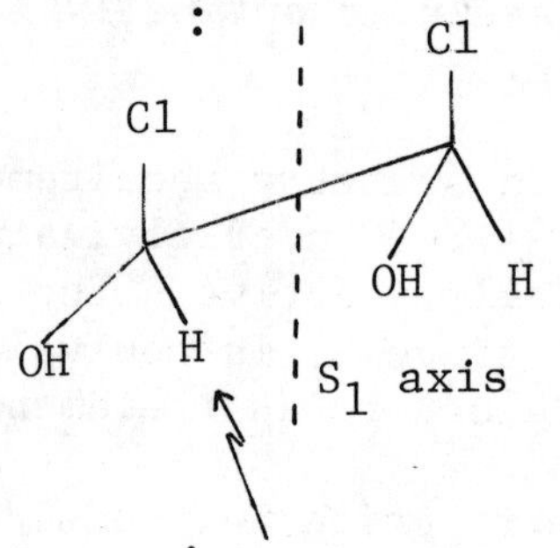

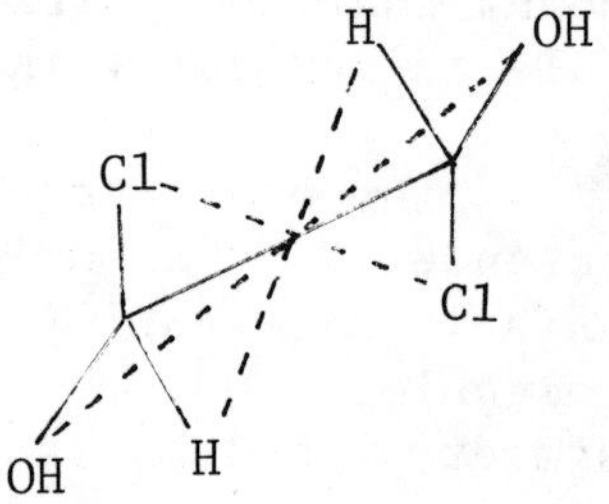

Most symmetric conformation.

Q-68 Which of the following does this molecule possess: S_n axis, C_n axis, chiral axis?

Is the molecule optically active? Explain

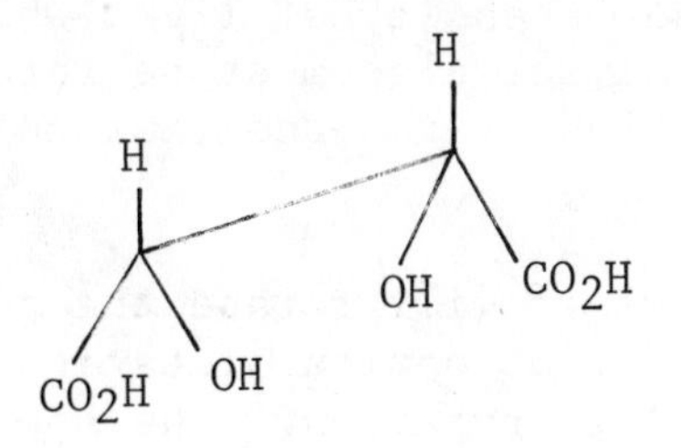

A-68 The molecule's symmetry is limited to a C_2 axis. (It possesses no S_n axis.) For this reason it is optically active. It has no chiral axis because it does not have the structure of an appropriately substituted, elongated tetrahedron.

Q-69 Which of the following does this molecule possess: S_n axis, C_n axis, chiral axis.

Is the molecule optically active? Explain.

A-69 The molecule's symmetry is limited to a C_2 axis. It possesses no S_n axis. For either of these reasons it is optically active. It has a chiral axis colinear with the C=C=C atoms.

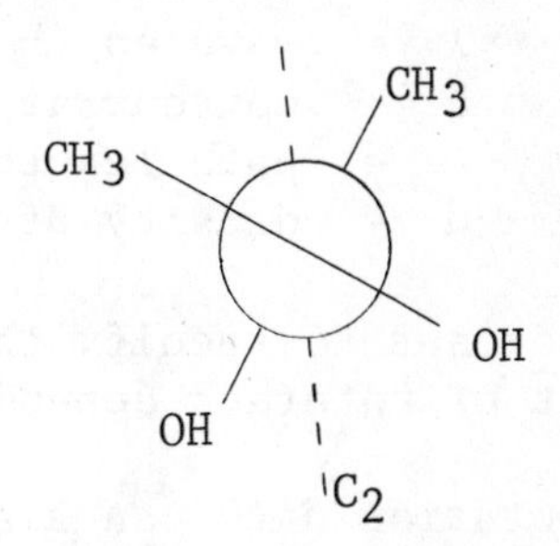

Optical Rotation:

S-7 The rotation of the plane of plane-polarized light by an optically active compound is measured in an instrument called a polarimeter.

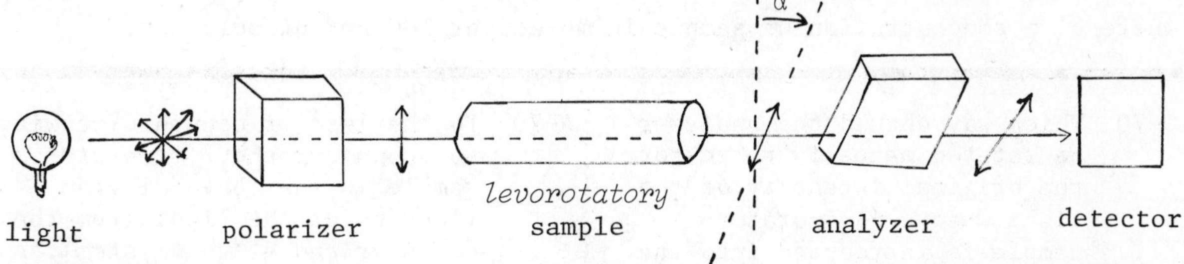

The monochromatic light source (usually the sodium D line, 589 nanometers) emits a beam of ordinary light whose components vibrate in all directions perpendicular to the line of propagation. A polarizer (Nicol prism or polaroid sheet) fairly sharply allows most of the light in one particular plane to pass through, removing the light in the other planes to produce plane-polarized light. An optically active sample placed in the beam of plane-polarized light will rotate the plane of polarization by an angle α, thus reducing the amount of light which passes through the analyzer (also a Nicol prism or polaroid sheet) which is initially arranged so only light in the original plane will pass through. To restore the intensity of light which passes through the analyzer to a maximum, the analyzer also must be rotated through angle α, so that the plane of polarization of the analyzer coincides with the plane of the rotated light.

The longer or more concentrated a sample, the more it will rotate the plane of polarized light. In order to facilitate comparison of optical rotation measurements, data are usually expressed as specific rotation. The specific rotation is defined as follows:

$$[\alpha]_{\lambda}^{T} = \frac{\alpha}{lc} \quad \text{for solutions}$$

$$[\alpha]_{\lambda}^{T} = \frac{\alpha}{ld} \quad \text{for neat (pure, no solvent) liquids}$$

where α = measured angle of rotation
T = temperature
λ = wavelength of light in nanometers (nm)
c = concentration of chiral substance in g per cm^3 of solution
l = path length in decimeters (dm, 1 dm = 10 cm)
d = density of neat liquid (no solvent) in g per cm^3

It is important to specify the temperature, solvent, and wavelength since the extent of rotation depends on these factors.

Optical rotation data can also be expressed as molecular rotation

$[M]$ or $[\Phi]$, which is defined as follows:

$$[M]_{\alpha}^{T} = [\Phi]_{\lambda}^{T} = \frac{\alpha}{lc'} = \frac{[\alpha]_{\lambda}^{T}\ (\text{mol wt})}{100}$$

where c' = concentration of sample in moles per 100 cm^3 of solution.

Q-70 Which way should the analyzer be rotated manually to restore the original intensity of the light when a levorotatory sample is introduced into the path of the light? Employ a rotation direction which gives a minimum angle of rotation, as shown in the above diagram.

A-70 To the left or counterclockwise, as shown for the *levorotatory* sample in the above diagram (looking at the light from the detector end of the system). A sample is called *levorotatory* when it rotates light in this manner.

Q-71 One could also restore the intensity of the light by rotating the polarizer instead of the analyzer. Which way would the polarizer be rotated for a levorotatory sample? Employ a rotation direction which gives a minimum angle of rotation.

A-71 The polarizer would be rotated to the right or clockwise (when viewed from the detector end of the system).

Q-72 What notation should be used to report a specific rotation which was measured at 288 K using the sodium D line?

A-72

$$[\alpha]_{D}^{288} \quad \text{or} \quad [\alpha]_{589}^{288}$$

Q-73 Express the following data in standard notation: a rotation of -76.3 = $\pm$ 0.3° was obtained for a chiral compound A (concentration = 1 g of A per cm^3 of ethanol solution) in a 1 dm long tube at 298 K using the D line of a sodium lamp.

A-73

$$[\alpha]_{D}^{298} = -76.3 \pm 0.3$$

(compound A, 1 g/cm^3, ethanol)

The negative sign of the rotation indicates that the rotation is to the left, that is, that the sample is *levorotatory*.

Q-74 Express the data below in standard notation: a pure sample of compound B in a 1 dm tube at 293 K rotated the plane of a beam of plane-polarized light of wavelength 300 nm by 38 $\pm$ 0.5 to the left.

A-74

$$[\alpha]_{300}^{293} = -38 \pm 0.5$$

(compound B, neat)

Q-75 Express the data below in standard notation: 10 g of compound C in 50 cm^3 of aqueous solution rotated the plane-polarized D line of a sodium lamp by 10 $\pm$ 0.2° *dextrorotatory* at 298 K in a 20 cm long tube. (The rotation is positive when the sample rotates the light to the right. Such a sample is called *dextrorotatory*.)

A-75

$$[\alpha]_{D}^{298} = +25 \pm 0.2^{\circ}$$

(compound C, 0.2 g/cm^3, H_2O)

Note: $[\alpha]_{D}^{298} = \frac{\alpha}{lc}$

$\alpha = +25 \pm 0.2^{\circ}$
$l = 2$ dm
$c = 10/50$ g/cm^3

Enantiomers:

S-8 Enantiomers are stereoisomers which are mirror images of each other but are not superimposable. The term applies to objects as well as molecules. Since a person's hands, if perfectly shaped, are nonsuperimposable mirror images of each other, then they are enantiomers of each other.

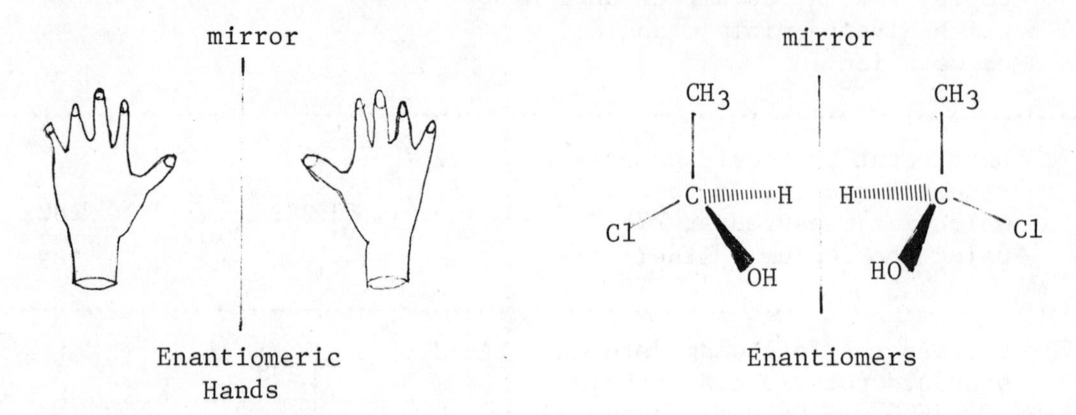

Enantiomeric Hands

Enantiomers

Q-76 How are a person's hands similar to a pair of enantiomers? Compare hands and enantiomers in terms of (1) the distance between the corresponding component parts in each member of the pair, and (2) their interaction with symmetric objects.

A-76 Just as the distances between any two corresponding fingers on each of a person's hands are the same, the distances between any two corresponding atoms in each of a pair of enantiomeric molecules are the same. Just as each hand will interact in the same way with symmetric objects like spoons or hammers, enantiomeric molecules react identically with symmetric agents. For instance, enantiomers are equally reactive towards symmetrical chemical reagents. Enantiomers also behave equivalently toward "symmetrical" physical changes like temperature changes (they have the same melting and boiling points) and solution composition changes (they have the same solubilities). Enantiomers are identical in all chemical and physical properties toward symmetric agents and conditions.

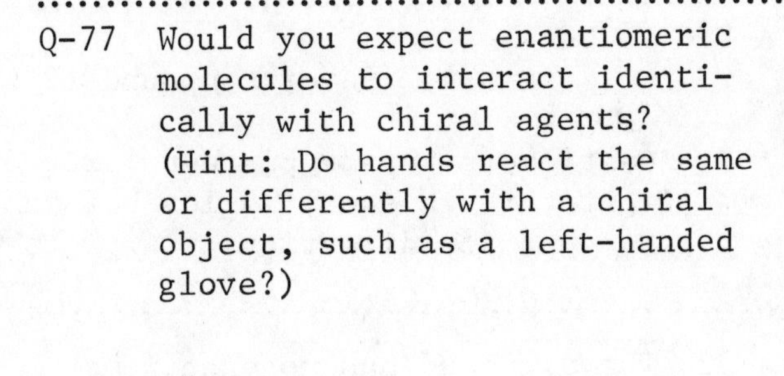

Q-77 Would you expect enantiomeric molecules to interact identically with chiral agents? (Hint: Do hands react the same or differently with a chiral object, such as a left-handed glove?)

A-77 Just as right and left hands react differently with a chiral object, such as a left-handed glove, a given enantiomeric molecule will interact differently with chiral agents. For example, enantiomers interact differently with plane-polarized light and with chiral chemical reagents.

Q-78 If one enantiomer rotates plane-polarized light 160° in a clockwise direction, what will the other enantiomer do?

A-78 Each enantiomer will rotate plane-polarized light through the same angle. The two enantiomers, however, will rotate light in opposite directions. If one rotates the light 160° in a clockwise direction, the other will rotate light 160° in a counterclockwise direction. The one which rotates light in a clockwise direction is called *dextrorotatory* (*dextro* or *d* or +) while the other is termed *levorotatory* (*levo* or *l* or -).

Q-79 Draw the enantiomer of this stereoisomer of *l*-phenylethanol.

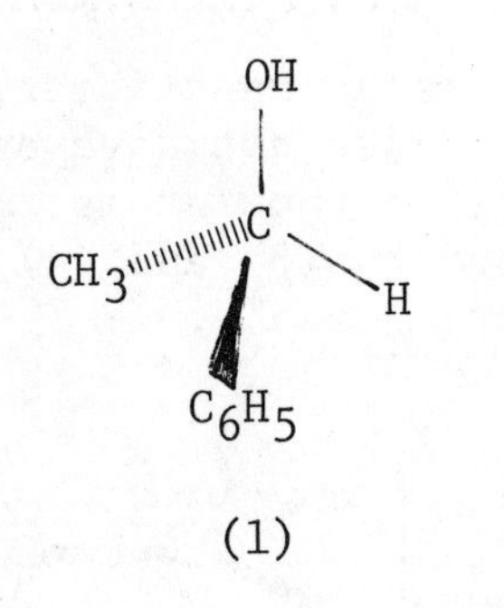

(1)

A-79

mirror

OH
C
H
CH_3
C_6H_5

(2)

Q-80 Which of the stereoisomers of *l*-phenylethanol, (1) and (2), is optically active?

A-80 Since they are enantiomers, they are both optically active.

Q-81 If $[\alpha]^{300}_{D}$ = + 42.9 (neat) for compound (1), then what is the specific rotation for compound (2)?

A-81 $[\alpha]^{300}_{D}$ = − 42.9 for compound (2).

Enantiomers rotate plane-polarized light through equal angles but in the opposite direction.

Q-82 Draw the enantiomer of this compound.

CH_3
H C H
Cl

A-82 The compound has no enantiomer because it is superimposable on its mirror image. The center carbon is not asymmetric because only three of the four groups are different.

Q-83 Does this compound have an enantiomer? Explain. Is it asymmetric?

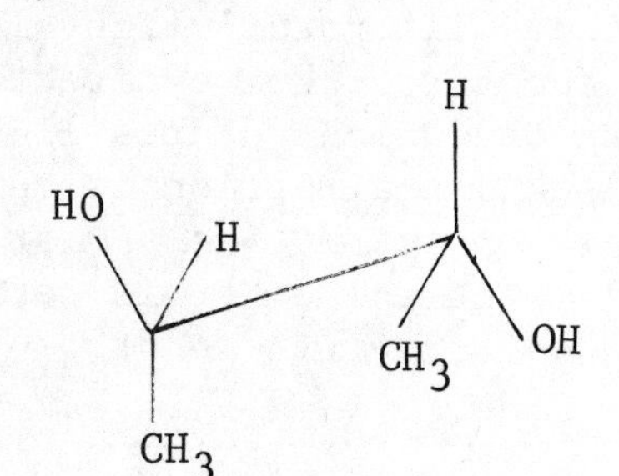

A-83 Yes, it has an enantiomer for two equivalent reasons:

(1) It is not superimposable on its mirror image and is thus chiral.

(2) It lacks an S_n axis.

Note: It is not asymmetric. This is because it has a C_2 axis (when drawn in a more symmetrical conformation). See Q-52, p. 53.

Q-84 Does this compound have an enantiomer? If so, draw it.

CH_3
C=C=C H
H CH_3

1,3-dimethylallene

(1)

Label the asymmetric carbon atom.

A-84 Yes, it has an enantiomer, although it does not have any asymmetric carbon atoms. Stereoisomers (3) and (4) are not superimposable.

CH_3 C=C=C H CH_3 H | H C=C=C CH_3 CH_3 H

(3) (4)

mirror

Q-85 Will these compounds rotate plane-polarized light in the same or opposite directions?

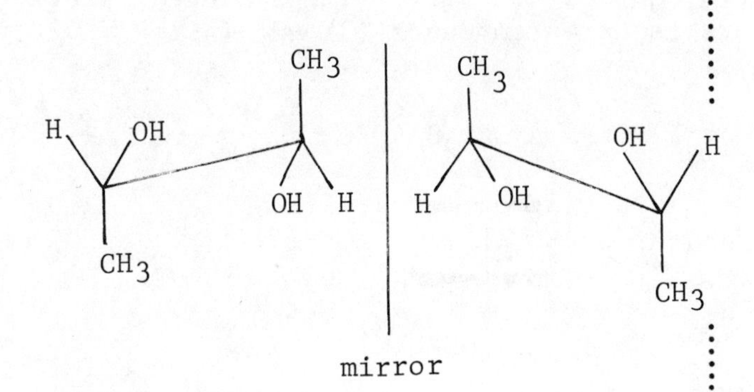

A-85 Neither, the structures are mirror images but they are superimposable. In other words, they are identical. The single compound they represent is optically inactive because it has a center of symmetry (*i*, same as S_2).

Q-86 Does *trans*-cyclooctene have an enantiomer? If so, draw it.

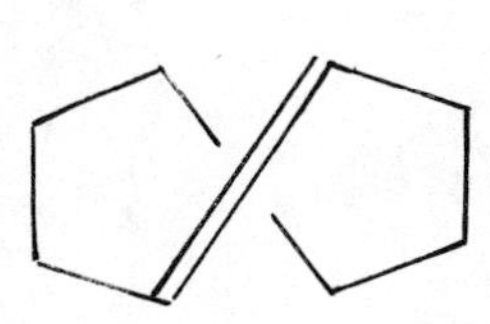

trans-cyclooctene

A-86 Yes, here it is.

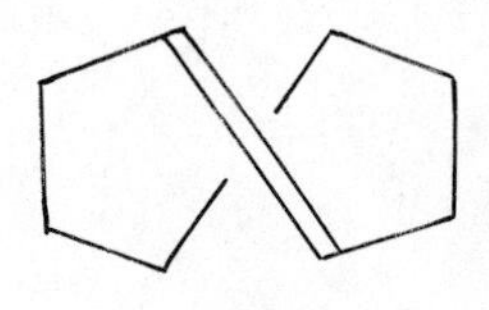

Q-87 Does this compound have an enantiomer? If so, draw it.

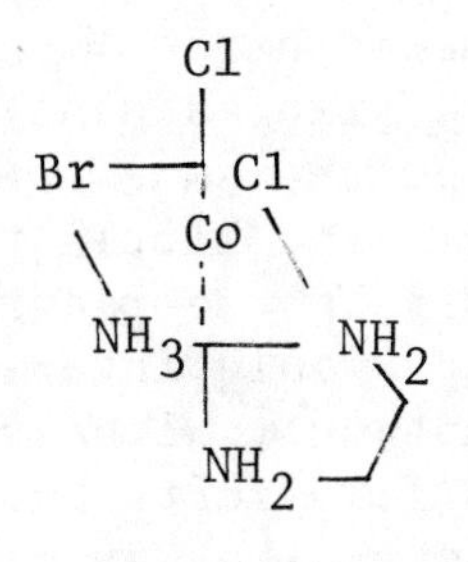

A-87 Yes, here it is.

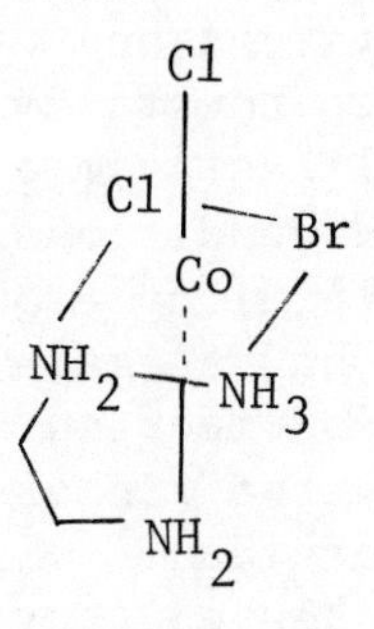

Diastereomers and Meso Compounds:

S-9 Diastereomers are stereoisomers which are not enantiomers of each other. Diastereomers are therefore non-superimposable, non-mirror image stereoisomers. Compounds (1) and (2) are diastereomers; so are compounds (3) and (4).

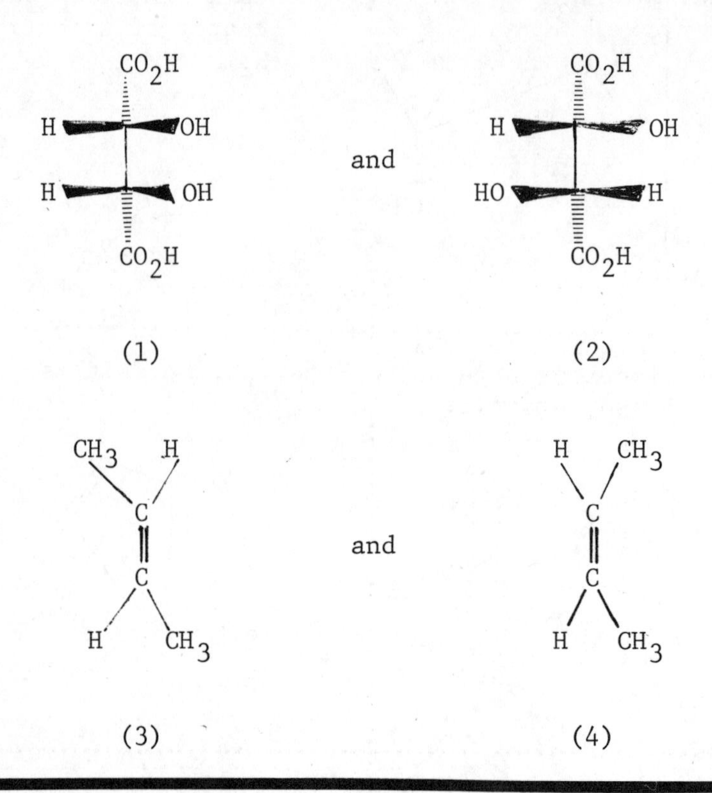

Q-88 In contrast to enantiomers, diastereomers have different physical and chemical properties toward symmetrical as well as asymmetrical reagents and conditions. It may be helpful to employ a comparison of hands again to elaborate this important comment. If any two fingers on one of an individual's hands could be exchanged (say the thumb and the middle finger on one hand were switched) then his hands would become like diastereomers of each other. Explain.

A-88 The particular distances between fingers and the positions of the fingers on each of these diastereomeric hands (a normal hand and a rearranged hand) would no longer be the same. For instance, the thumb and little finger would be closer in the rearranged hand. The hands would interact differently not only with asymmetric objects like left- and right-hand gloves, but even with symmetric objects like spoons or hammers. Similarly, diastereomers have different chemical and physical properties toward symmetric as well as asymmetric reagents and conditions.

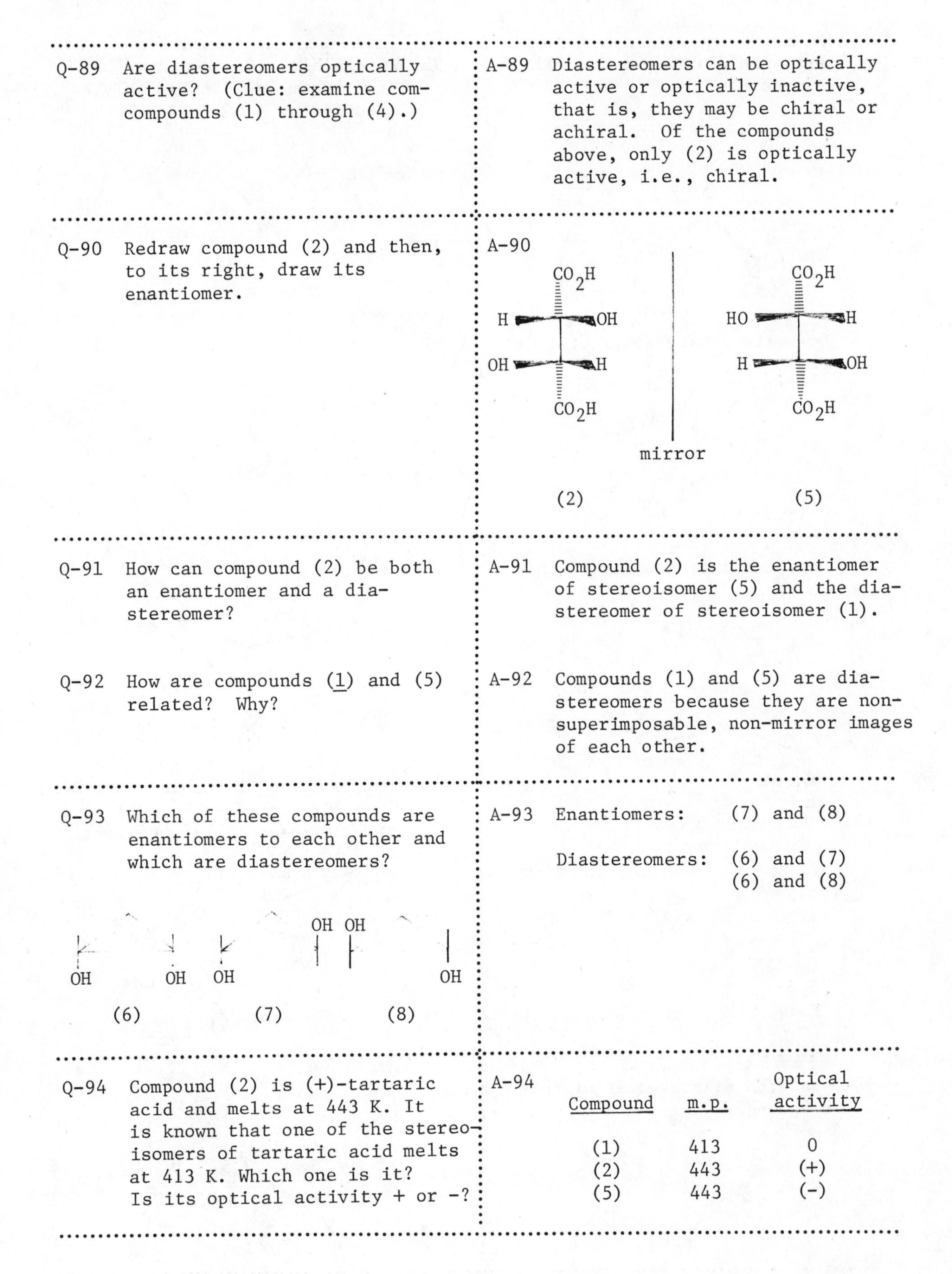

Q-89 Are diastereomers optically active? (Clue: examine com- compounds (1) through (4).)

A-89 Diastereomers can be optically active or optically inactive, that is, they may be chiral or achiral. Of the compounds above, only (2) is optically active, i.e., chiral.

Q-90 Redraw compound (2) and then, to its right, draw its enantiomer.

A-90

	CO_2H			CO_2H	
H		OH		HO		H
OH		H		H		OH
	CO_2H			CO_2H	

mirror

(2) (5)

Q-91 How can compound (2) be both an enantiomer and a diastereomer?

A-91 Compound (2) is the enantiomer of stereoisomer (5) and the diastereomer of stereoisomer (1).

Q-92 How are compounds (1) and (5) related? Why?

A-92 Compounds (1) and (5) are diastereomers because they are non-superimposable, non-mirror images of each other.

Q-93 Which of these compounds are enantiomers to each other and which are diastereomers?

OH OH OH OH OH OH

(6) (7) (8)

A-93 Enantiomers: (7) and (8)

Diastereomers: (6) and (7)
(6) and (8)

Q-94 Compound (2) is (+)-tartaric acid and melts at 443 K. It is known that one of the stereoisomers of tartaric acid melts at 413 K. Which one is it? Is its optical activity + or -?

A-94

Compound	m.p.	Optical activity
(1)	413	0
(2)	443	(+)
(5)	443	(-)

Q-95 Draw the diastereomer of this compound.

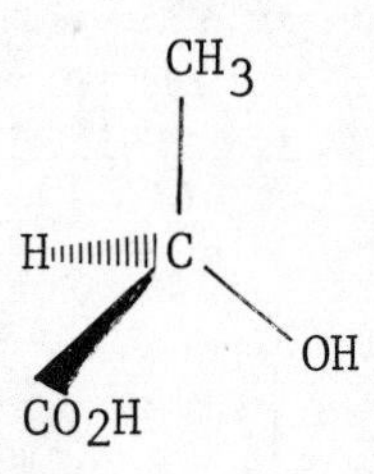

A-95 Impossible, this compound has no diastereomers. It has only an enantiomer.

Q-96 Draw the diastereomer of this compound.

CH_3

H

A-96

H

CH_3

Q-97 Draw the diastereomer of this compound.

CH_3

H

H

CH_3

(9)

A-97

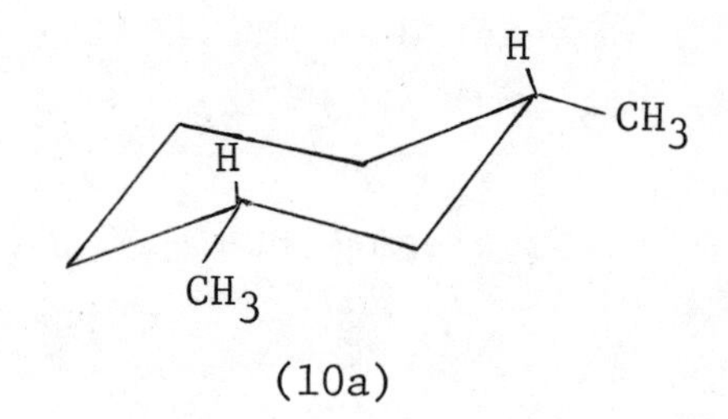

(10a)

This is the most stable conformer of the diastereomer. (10b) is a somewhat less stable conformer.

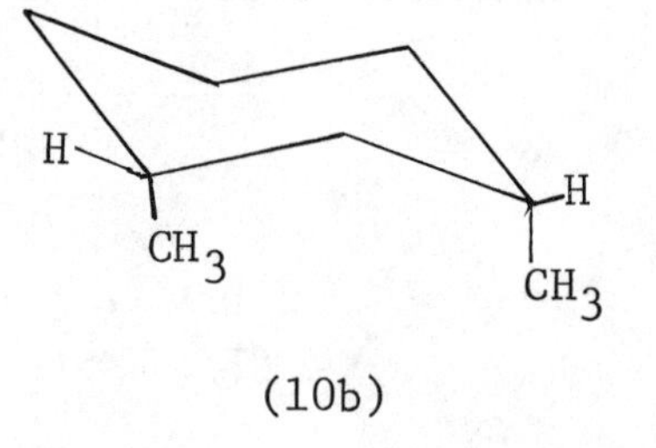

(10b)

Q-98 Draw the diastereomer of this compound.

CH_3 Cl

C=C

H C_2H_5

A-98

CH_3 C_2H_5

C=C

H Cl

Q-99 Are these compounds diastereomers?

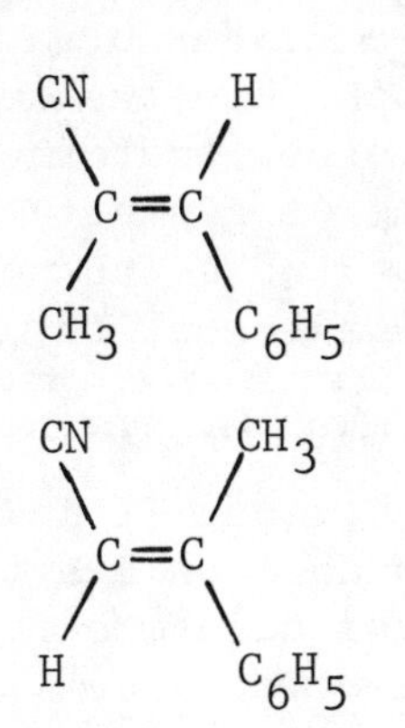

A-99 No, they are position isomers. In each compound the groups bonded to a given carbon atom are different.

Q-100 A compound which contains more than one chiral element (especially asymmetric carbon atoms) and yet is superimposable on its mirror image is a *meso* compound. Are *meso* compounds optically active?

A-100 *Meso* compounds are optically inactive because they are superimposable on their mirror images. They therefore are achiral and must possess one or more symmetry elements besides a C_n axis.

Q-101 What features make this a *meso* compound? Explain.

H, OH, H, OH, CO_2H, CO_2H

A-101 It has two chiral centers and yet it possesses a plane of symmetry.

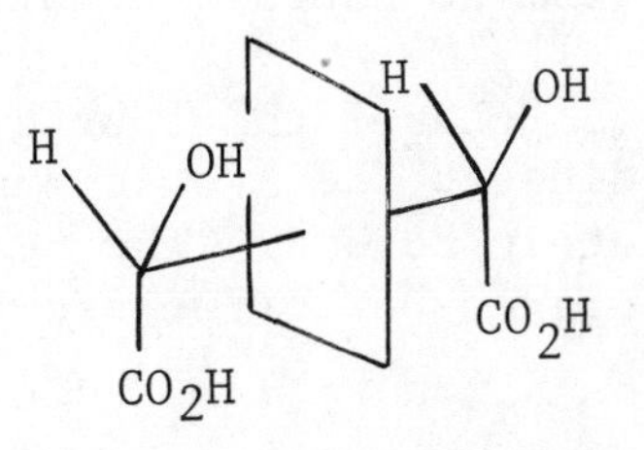

Q-102 Is this a *meso* compound? Explain. (Hint: It has a plane of symmetry.)

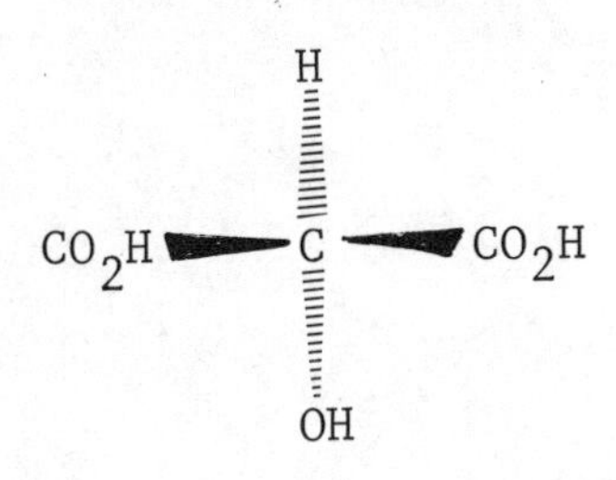

A-102 No, because it has no chiral elements.

Q-103 Is this a *meso* compound? Explain.

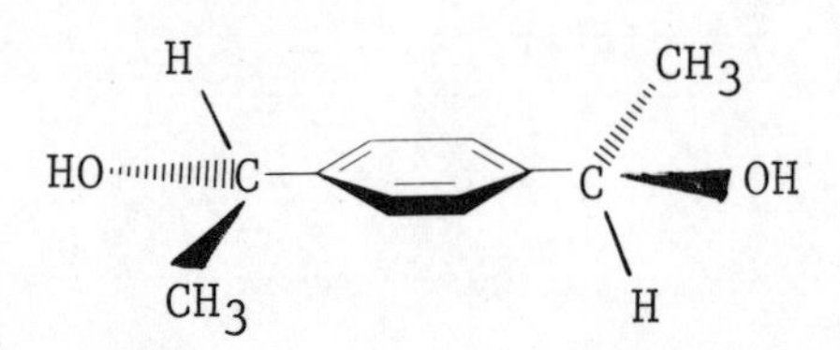

A-103 Yes, it has two chiral centers (two asymmetric carbons) and has a center of symmetry in the conformer shown. One of the other conformers has a plane of symmetry. Either of these two conformers is superimposable on its mirror image.

Q-104 Is this a *meso* compound? Explain.

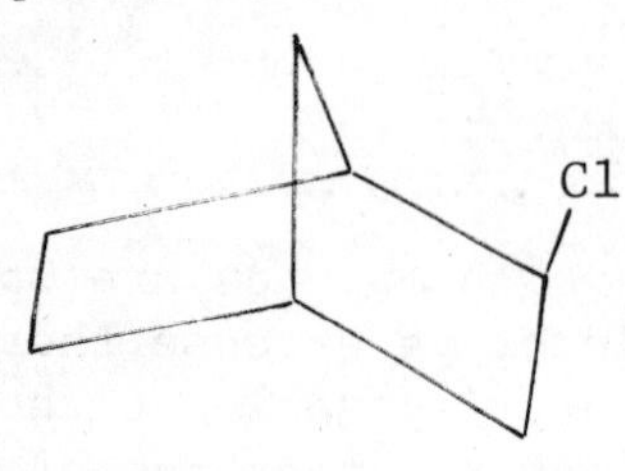

A-104 No, because it is not superimposable on its mirror image. It has no symmetry elements whatsoever.

The Sequence Rule I: Priority of Groups:

S-10 The Cahn, Ingold, Prelog sequence rules provide a way to specify the absolute molecular chirality of a compound. Each chiral element in the molecule is assigned either the symbol R or S.

To determine whether a chiral center (an asymmetric carbon atom) is R or S:

(1) Label the four groups "a,b,c,d," in order of decreasing priority according to the atomic numbers of the atoms attached directly to the asymmetric carbon. The heaviest group is assigned "a," the lightest group is assigned "d." Two examples are shown.

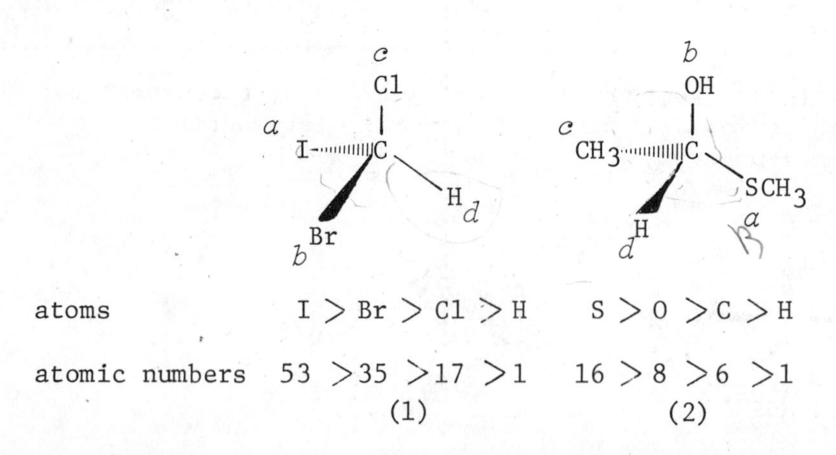

(2) Draw or view the model so that the lightest ligand *d* points directly back. Determine whether tracing a path from groups *a* to *b* to *c* gives a clockwise (R, Latin *rectus*, for right) or a counterclockwise course (S, Latin *sinister*, for left). This constitutes the chirality assignment.

When structures (1) and (2) are redrawn so that the *d* group in each is in the back, structures (3) and (4) are obtained, respectively. Therefore compound (1) is (S) and compound (2) is (R).

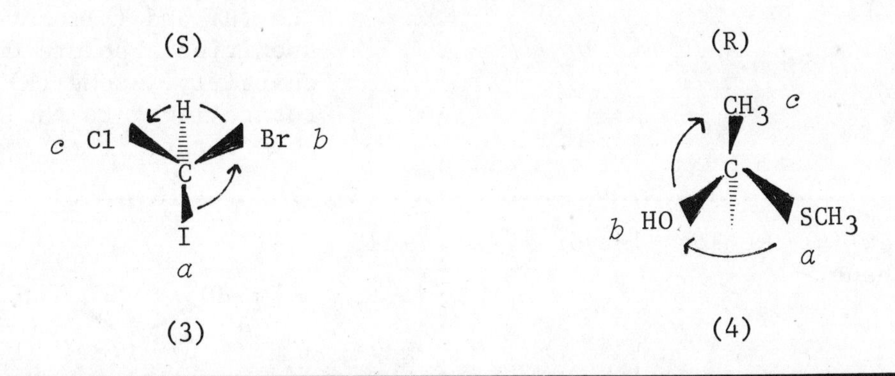

Q-105 Arrange these atoms in order of decreasing priority:

C, Br, O, S

Which one would be labeled *b*?

A-105

Br > S > O > C
35 > 16 > 8 > 6

The S would be labeled *b*.

Q-106 Is this molecule (R) or (S)? (Clue: Assign the groups a, b, c, d and then redraw the structure placing the group labeled *d* in the back.)

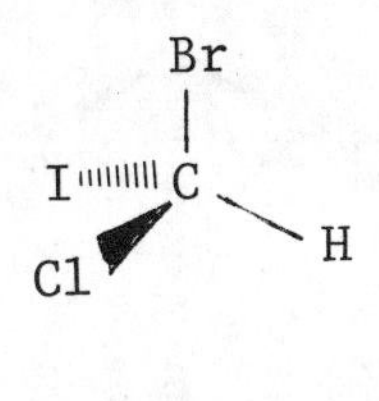

(5)

A-106 It is (R).

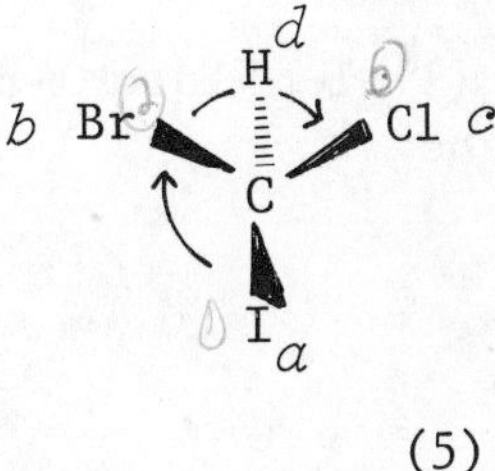

Q-107 Is this molecule (R) or (S)?

A-107 It is (S). It is the same as compound (1).

Q-108 How are compounds (1) and (5) related?

A-108 They are enantiomers: (1) is the (S) enantiomer and (5) is the (R) enantiomer.

Q-109 Is an (R) enantiomer dextrorotatory or levorotatory?

A-109 It may be either. The (R) and (S) convention has nothing to do with the direction of rotation of plane-polarized light. The (R) and (S) convention specifies absolute molecular chirality. Some (R) enantiomers rotate light to the left and some rotate it to the right.

Q-110 Draw (R)-1-chloro-1-hydroxyethane.

A-110

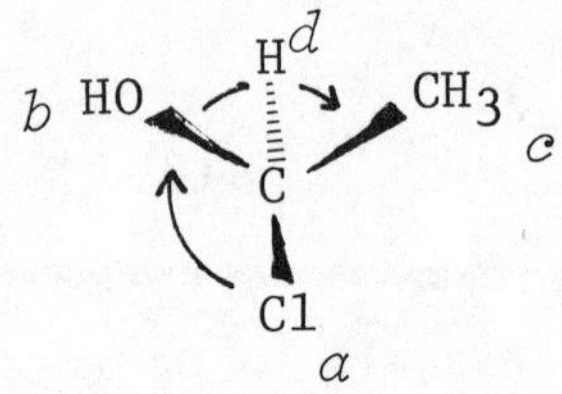

Q-111 Draw (S)-1-chloro-2-hydroxyethane.

A-111 Impossible, because the compound has no chiral centers.

$ClCH_2CH_2OH$

Q-112 Draw (S)-1-chloro-1-mercaptopropane.

A-112

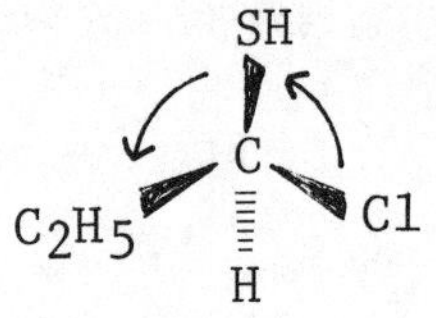

Cl > S > C
17 > 16 > 6

The Sequence Rule II: Secondary Priority:

S-11 Often two or more of the groups are attached to the chiral center through identical atoms. For instance, both the CH_3 and CH_2OH groups in (6) are joined through a carbon atom. In cases such as this, priority of groups (a > b > c > d) is established by the next atoms in the groups. If these atoms are still the same, then priority is established by working outward concurrently along these groups, atom by atom, up to the first point of difference. Priority is then determined by which atoms are attached at this point. Consider (6):

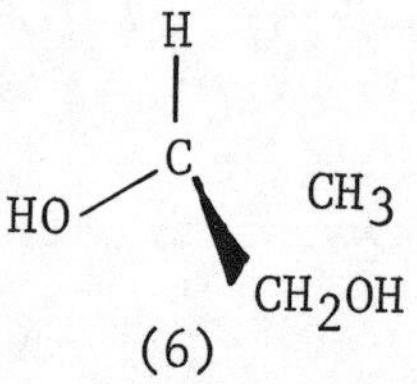

(6)

The priorities for the groups in this compound are determined as follows:

Group	First Atomic Number (atom)	Sum of Secondary Atomic Numbers (atoms)
OH	8 (O)	1 (H)
CH_3	6 (C)	3 (H + H + H)
CH_2OH	6 (C)	10 (O + H + H)
H	1 (H)	

Based on the first atomic number, OH has priority over the three other groups. Based on the secondary atoms, the CH_2OH group has priority over the CH_3 group. Therefore,

$$OH > CH_2OH > CH_3 > H$$

$$a > b > c > d$$

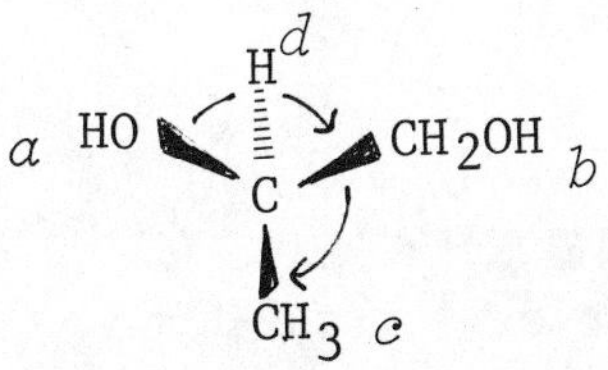

H *d*
a HO — C — CH_2OH *b*
CH_3 *c*

The compound is (R).

Q-113 Is this structure (R) or (S)?

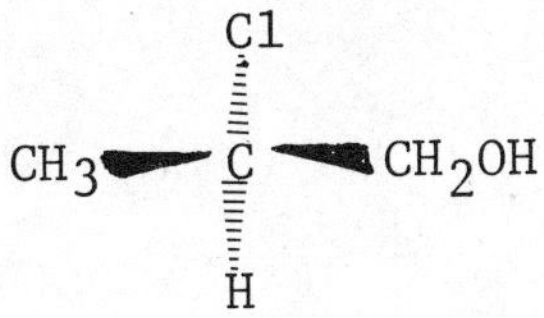

Cl
CH_3 — C — CH_2OH
H

A-113 It is (R).

Cl > C > H

C (HHO) > C (HHH)

Thus:

Cl > CH_2OH > CH_3 > H

a b c d

Q-114 Is this structure (R) or (S)?

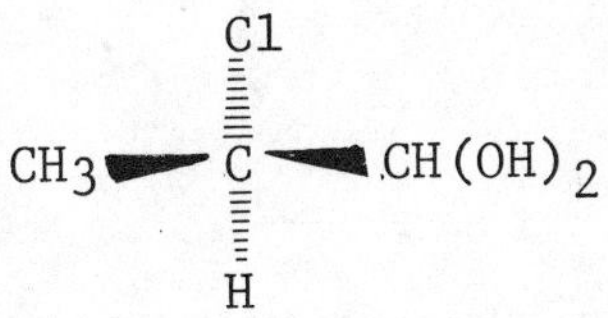

Cl
CH_3 — C — $CH(OH)_2$
H

A-114 It is (R). Even though the $CH(OH)_2$ group weighs more in terms of atomic numbers than the Cl atom, the Cl atom is assigned a higher priority because the decision involves Cl versus only the C atom of the group $CH(OH)_2$. Therefore,

Cl > $CH(OH)_2$ > CH_3 > H

a b c d

Q-115 Is this structure (R) or (S)?

$$\begin{array}{c} C(CH_3)_3 \\ | \\ CH_3\text{''''''}C\text{---}H \\ \blacktriangleleft \\ CH_2OH \end{array}$$

A-115 It is (R).

Group	First Atomic Number	Sum of Secondary Atomic Numbers
$C(CH_3)_3$	6 (C)	18 (C + C + C)
CH_2OH	6 (C)	10 (O + H + H)
CH_3	6 (C)	3 (H + H + H)
H	1 (H)	

$\therefore C(CH_3)_3 > CH_2OH > CH_3 > H$

a b c d

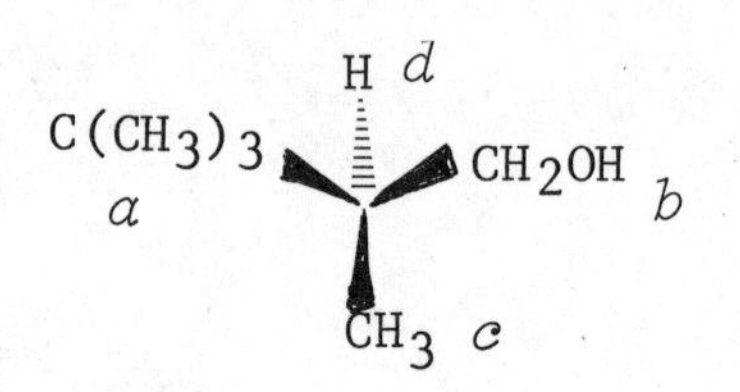

Q-116 Draw the (S) enantiomer of this compound.

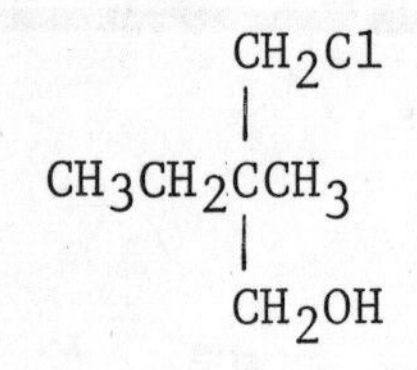

A-116

Group	First Atomic Number	Sum of Secondary Atomic Numbers
CH_2Cl	C (6)	19 (H + H + Cl)
CH_2OH	C (6)	10 (H + H + O)
CH_2CH_3	C (6)	8 (H + H + C)
CH_3	C (6)	3 (H + H + H)

$\therefore CH_2Cl > CH_2OH > CH_2CH_3 > CH_3$

a b c d

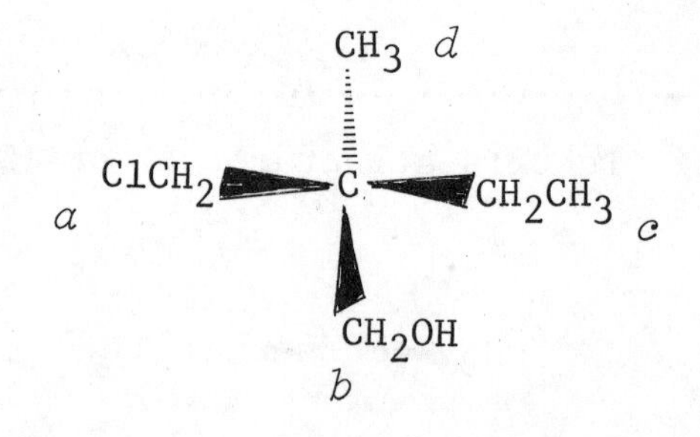

Q-117 Draw the (R) enantiomer of this compound.

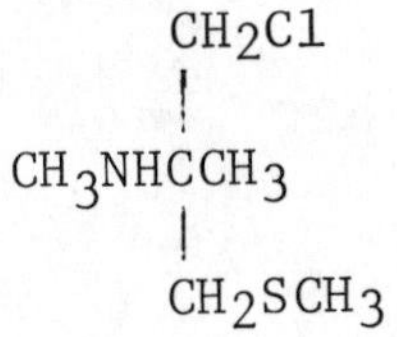

$$CH_3NHC(CH_2Cl)(CH_2SCH_3)CH_3$$

A-117

Group	First Atomic Number	Sum of Secondary Atomic Numbers
$NHCH_3$	7 (N)	
CH_2Cl	6 (C)	19 (H + H + Cl)
CH_2SCH_3	6 (C)	18 (H + H + S)
CH_3	6 (C)	3 (H + H + H)

$\therefore NHCH_3 > CH_2Cl > CH_2SCH_3 > CH_3$

a *b* *c* *d*

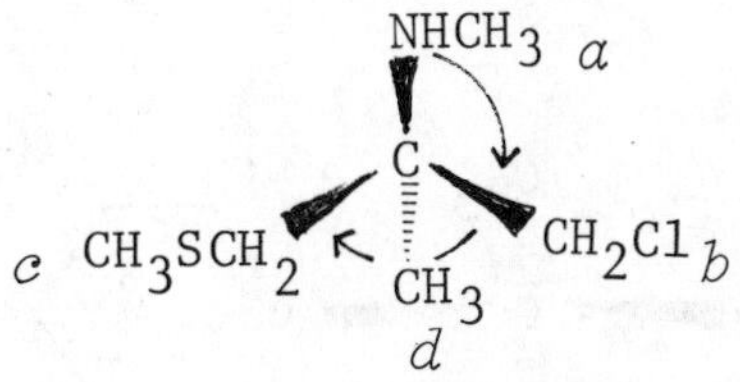

$NHCH_3$ *a*, CH_2Cl *b*, CH_3SCH_2 *c*, CH_3 *d*

The Sequence Rule III: Branch Points:

S-12 Sometimes groups are equivalent even at a branch point (branch points are starred in the examples below). When branch points are reached where the attached atoms are the same, one proceeds first along those branches which start with the heaviest atoms and then, if these branches are the same, one then proceeds along those branches which start with the next heaviest atoms until a point of difference is reached. Priority of groups is established based on the atoms at the first point of difference. In the following examples the arrows stop at the first point of difference.

Assignment

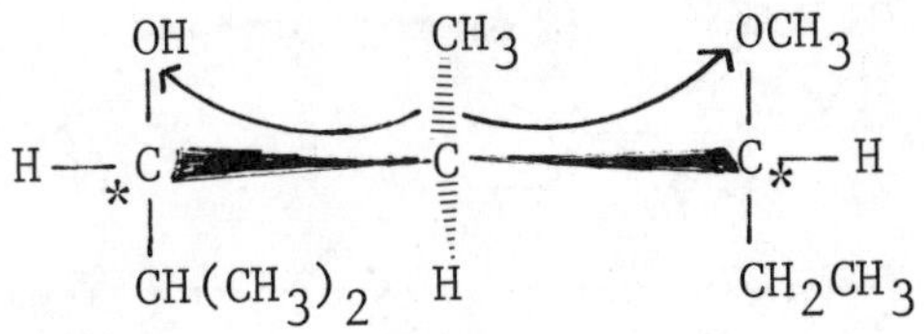

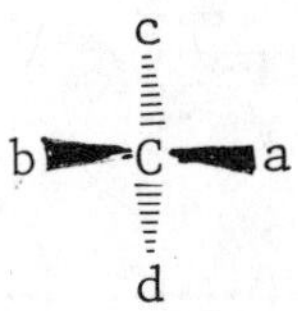

OH / H—*C / $CH(CH_3)_2$ — C (CH_3, H) — OCH_3 / *C—H / CH_2CH_3 b—C—a (c, d) (R)

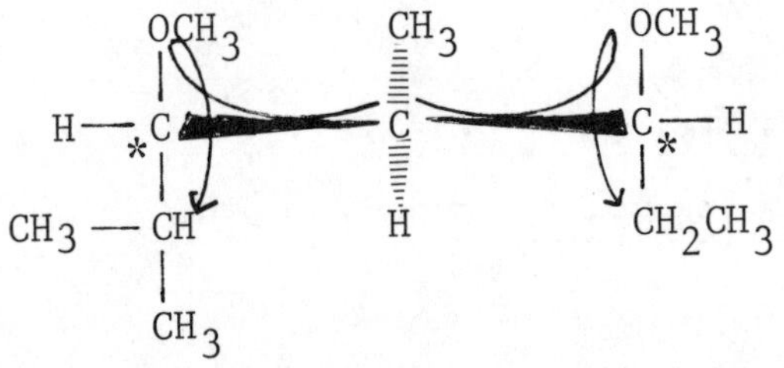

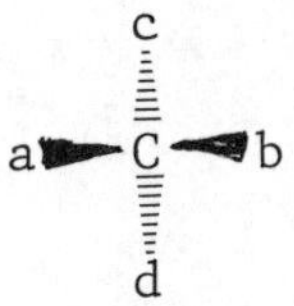

OCH_3 / H—*C / CH_3—CH—CH_3 — C (CH_3, H) — OCH_3 / *C—H / CH_2CH_3 a—C—b (c, d) (S)

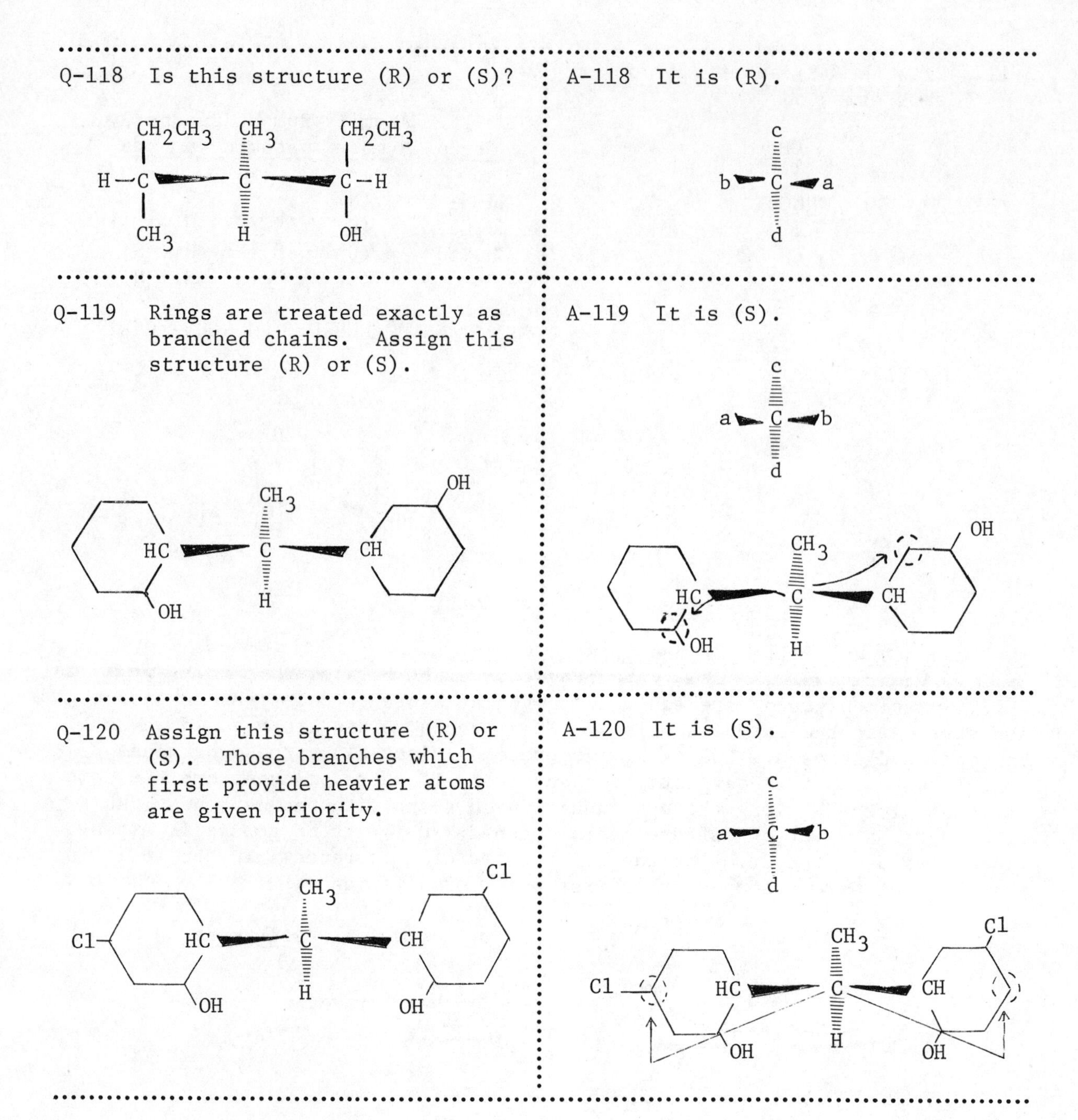

Q-118 Is this structure (R) or (S)?

A-118 It is (R).

Q-119 Rings are treated exactly as branched chains. Assign this structure (R) or (S).

A-119 It is (S).

Q-120 Assign this structure (R) or (S). Those branches which first provide heavier atoms are given priority.

A-120 It is (S).

The Sequence Rule IV: Multiple Linkages:

S-13 When two atoms are joined by a multiple bond, e.g. double, triple, both atoms are considered to be multiplied, e.g. doubled, tripled in determination of priority of groups.

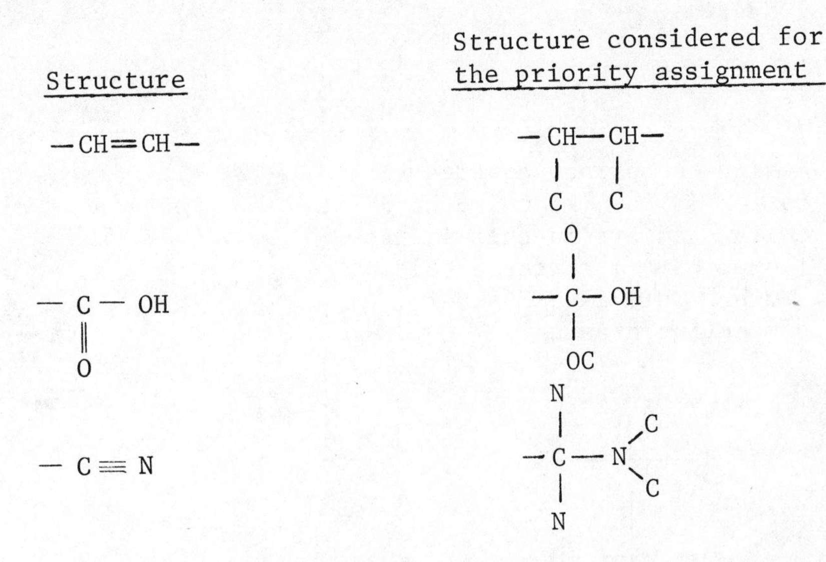

Therefore: $CO_2H > COCH_3 > CHO > CH_2OH$

$CN > C{\equiv}CH > CH{=}NH > CH{=}CH_2$

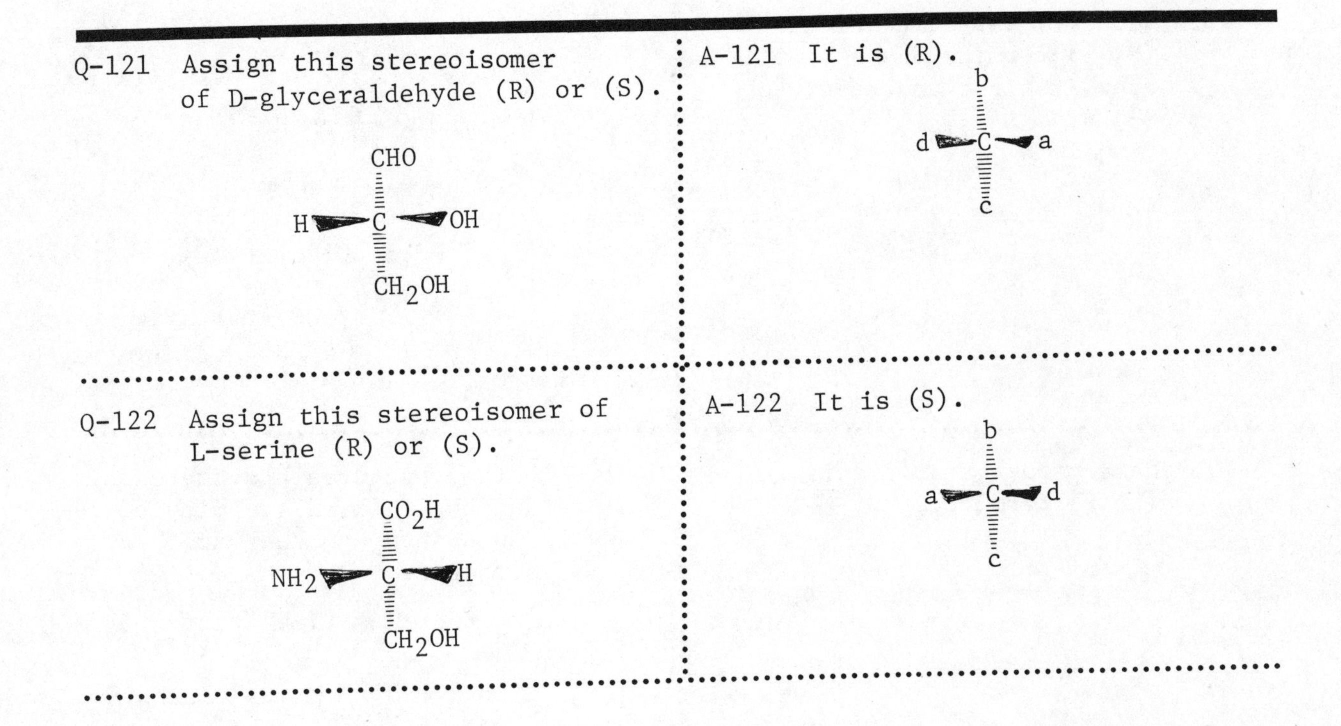

Q-121 Assign this stereoisomer of D-glyceraldehyde (R) or (S).

A-121 It is (R).

Q-122 Assign this stereoisomer of L-serine (R) or (S).

A-122 It is (S).

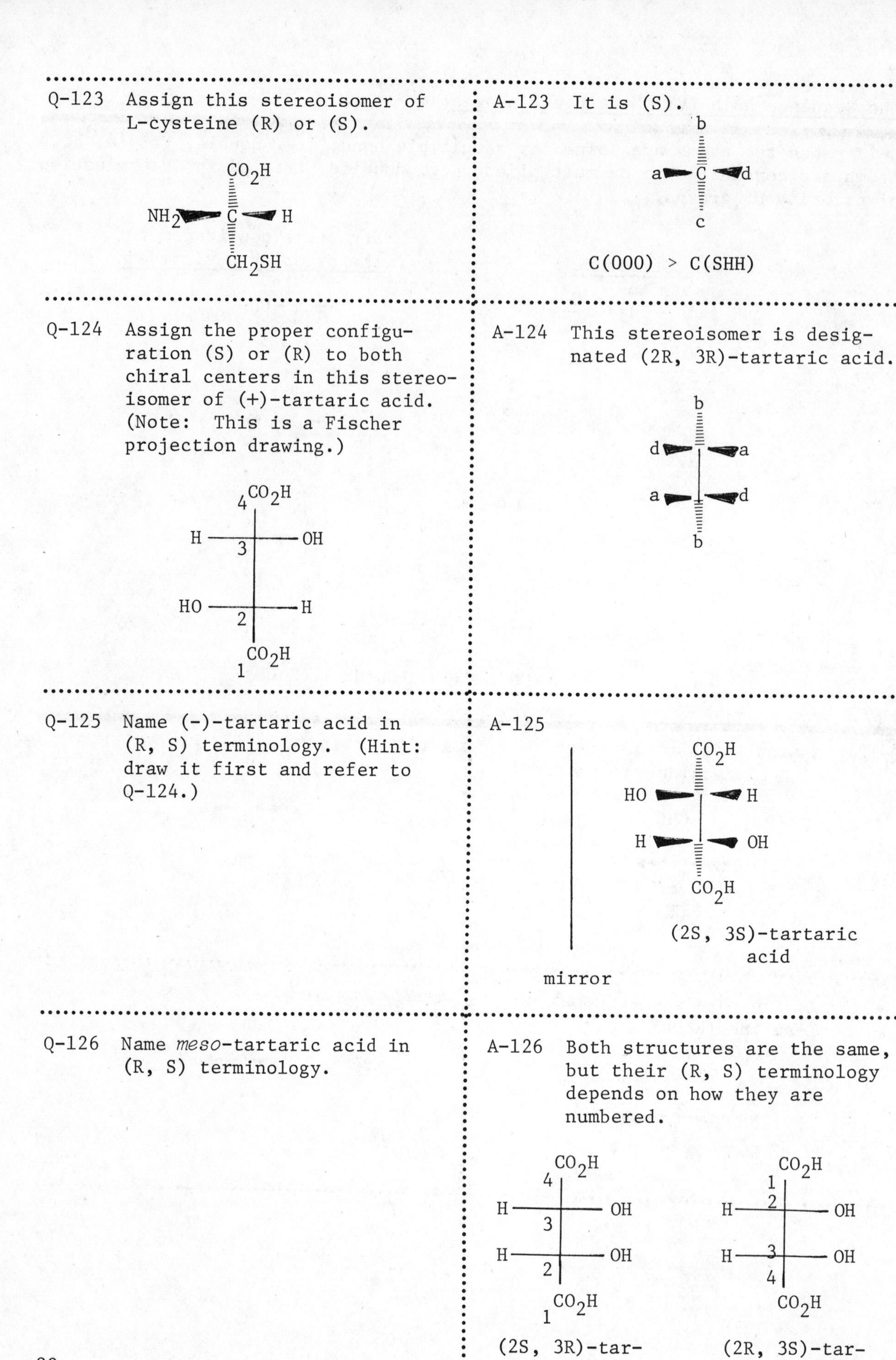

Q-123 Assign this stereoisomer of L-cysteine (R) or (S).

A-123 It is (S).

C(OOO) > C(SHH)

Q-124 Assign the proper configuration (S) or (R) to both chiral centers in this stereoisomer of (+)-tartaric acid. (Note: This is a Fischer projection drawing.)

A-124 This stereoisomer is designated (2R, 3R)-tartaric acid.

Q-125 Name (-)-tartaric acid in (R, S) terminology. (Hint: draw it first and refer to Q-124.)

A-125

(2S, 3S)-tartaric acid

Q-126 Name *meso*-tartaric acid in (R, S) terminology.

A-126 Both structures are the same, but their (R, S) terminology depends on how they are numbered.

(2S, 3R)-tartaric acid

(2R, 3S)-tartaric acid

S-14 For axial chirality, as illustrated by an elongated tetrahedron, it is not necessary that all four substituents be different. However, one must still be able to assign a unique priority to each substituent. Since it is immaterial from which end of the chiral axis** one views such a structure when assigning (R) or (S), it is required that "near groups" (with respect to the observer) are assigned a higher priority than "far groups." This distinguishes equivalent groups because when one of a pair of equivalent groups is near, the other one of the pair will always be far. If this is not the case, the structure is not chiral.

Assignment of axial chirality is carried out in two steps. First, one chooses one end of the chiral axis to look down to assign priorities $a > b > c > d$.

CH_3 / H — C=C=C — H / CH_3 ≡ a, b, c, d

(1)

Second, one views the model from the side opposite that of the lowest priority substituent, *d*, and traces a course from groups *a* to *b* to *c*. A clockwise course means the model is (R). A counterclockwise course means it is (S).

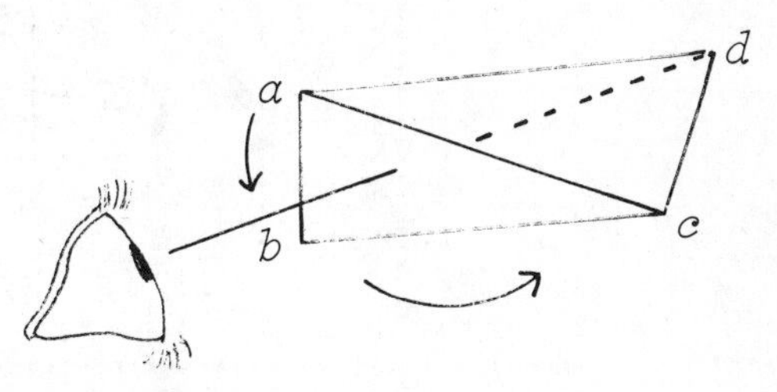

Therefore, compound (1) is (S).

Q-127 Look down the other end of the chiral axis of the allene (Compound 1, S-14) and determine whether the model is still (S).

A-127

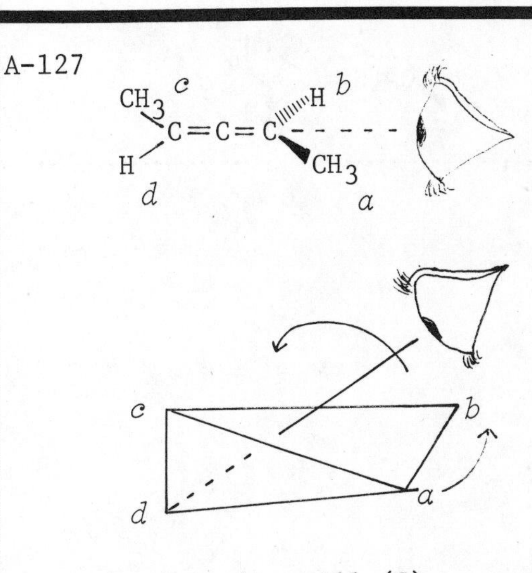

Yes, it is still (S).

**See page 54.

Q-128 Is this dichloroallene (R) or (S)?

A-128 It is (R).

Q-129 Substituted biphenyls are treated just like allenes. Is this structure (R) or (S)?

A-129 It is (R).

Q-130 Is this model (R) or (S)?

A-130 It has no (R) or (S) classification because it has no chiral elements. (See page 54.)

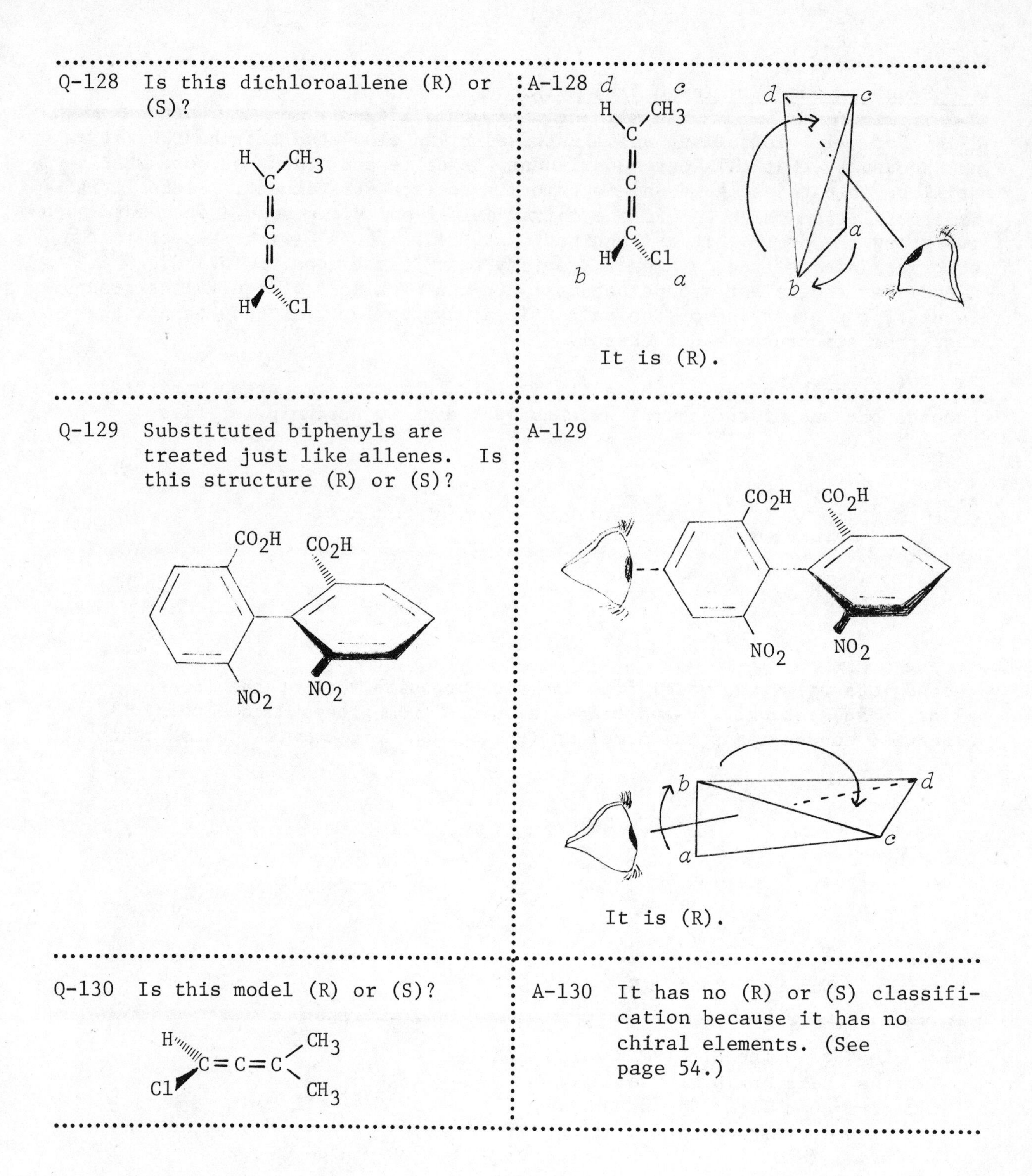

Prochirality:

S-15 A tetrahedrally bonded atom is a prochiral center when two and only two of its four ligands (atoms or groups) are exactly the same. The two equivalent ligands are "prochiral" because replacement of either one of them by a ligand different from the ones already present results in the "promotion" of the center to a chiral state.

prochiral | chiral

This is a non-trivial nomenclature because prochiral ligands can sometimes be distinguished both spectroscopically and enzymatically.

Q-131 Label the prochiral groups in this compound.

(1)

A-131

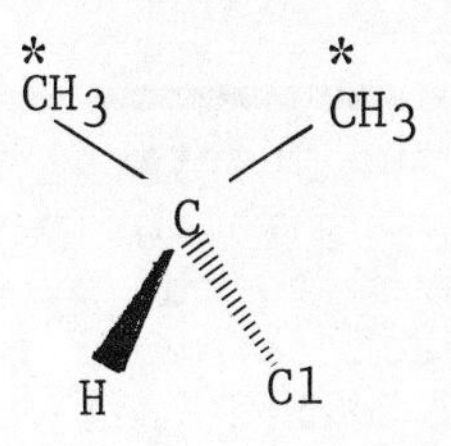

The two methyl groups are prochiral.

Q-132 Replace each of the methyl groups in (1) (separately) by Br to generate two new compounds. Determine the chirality (R or S) of each new compound.

A-132

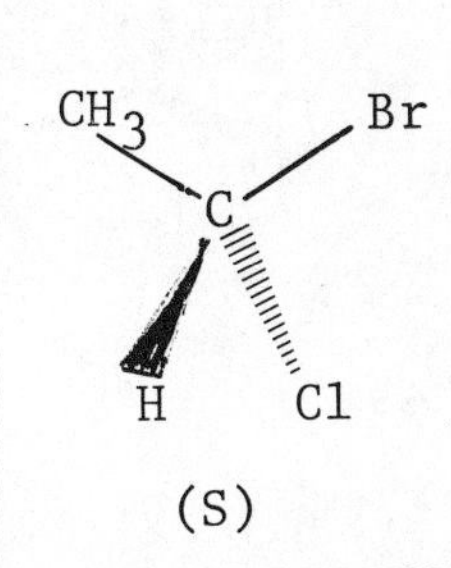

(S)

(R)

Q-133 Label the five prochiral centers in this compound

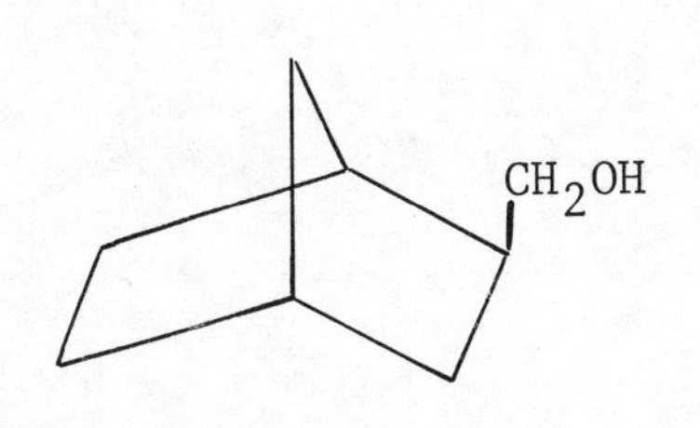

A-133

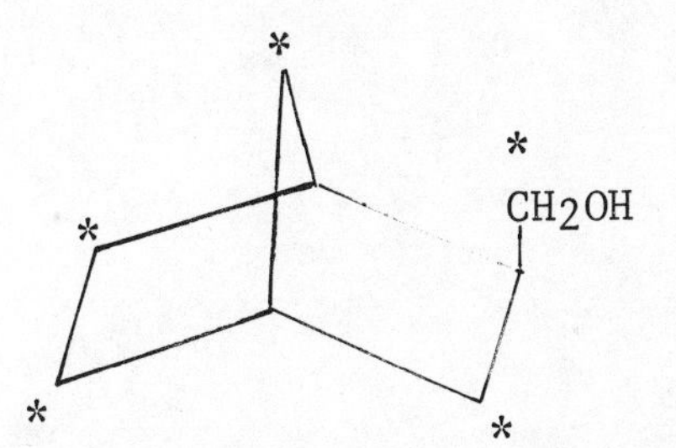

The three unlabeled carbons are chiral centers.

Q-134 Explain why this compound, shown in a Fischer projection, is or is not prochiral.

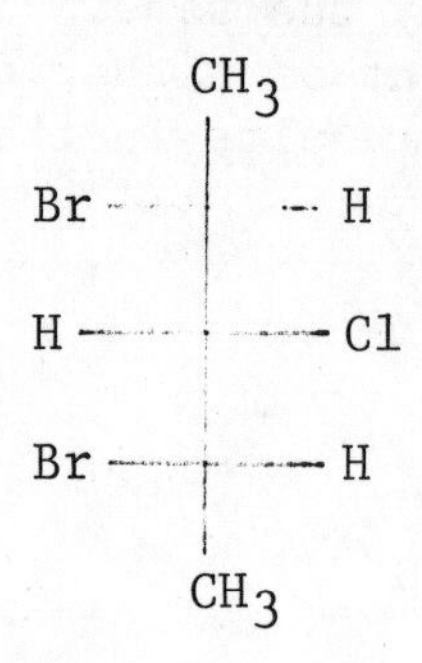

A-134 The compound is not prochiral. All of the tetrahedral centers have four different ligands. Even the two $CHBrCH_3$ groups on C_3 are not completely the same because they are enantiomeric, i.e. C_2 has R chirality and C_4 has S chirality.

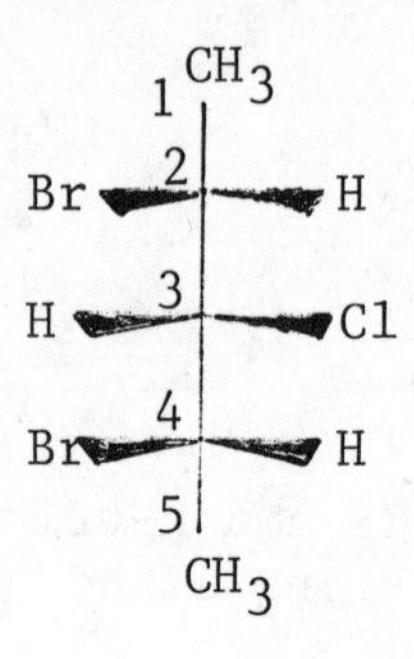

Pro-R, Pro-S:

S-16 Of the identical pair of atoms or groups in a prochiral compound, that atom or group which leads to an (R) compound when arbitrarily considered to be preferred to the other by the sequence rule (without change in priority with respect to other ligands) is termed "*pro*-R." The other identical ligand is termed "*pro*-S."

For instance, H_1 in compound (1) is *pro*-R because its arbitrary preference over H_2, as indicated by a, b, c, d in assignment (2), results in (R) chirality for the compound. Likewise, H_2 is *pro*-S because its arbitrary preference over H_1, as in assignment (3), results in (S) chirality.

(1) prochiral — (2) (R) chirality — (3) (S) chirality

Q-135 Determine which ligands in this compound are "*pro*-R" and which are "*pro*-S."

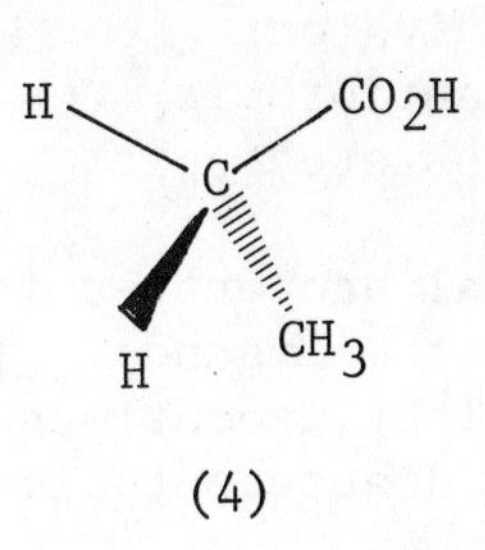

(4)

A-135

H_A is *pro*-S because assignment (5) is (S).

H_B is *pro*-R because assignment (6) is (R).

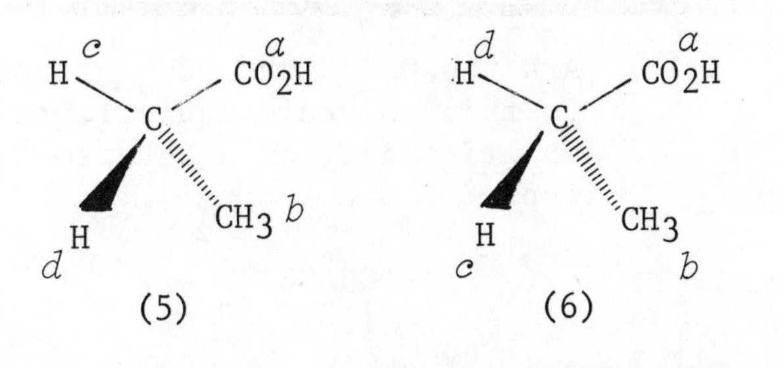

(5) (6)

Q-136 Which fluorine in this compound is *pro*-R?

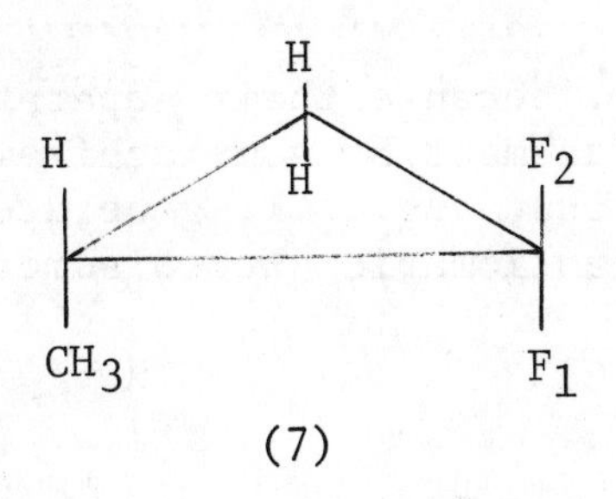

(7)

Atomic numbers: F (9), C (6).

A-136 F_2 is *pro*-R because assignment (8) is (R). (The CH_2 group is the least preferred ligand.)

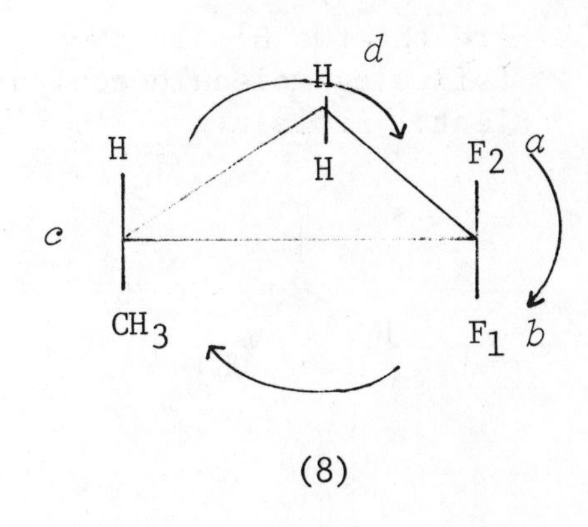

(8)

Equivalent, Enantiotopic, and Diastereotopic:

S-17 Often a molecule will contain two ligands (atoms or groups) composed of the same atoms or bonds. In many cases these ligands, while seemingly identical, are geometrically different. The two groups may be equivalent, enantiotopic or diastereotopic.

Equivalent ligands are not prochiral, which sets them apart from enantiotopic and diastereotopic ligands, both of which are prochiral.

One way to determine whether two ligands are equivalent, enantiotopic, or diastereotopic is to replace each of the two ligands by a ligand different from the ones already present:

If the replacement produces:	the ligands are:
superimposable compounds	equivalent
enantiomers	enantiotopic
diastereomers	diastereotopic

This classification according to stereochemical properties is useful in predicting the physical and chemical properties of the ligands. These properties will be the same if the ligands are equivalent, sometimes the same if they are enantiotopic, and always different for diastereotopic ligands.

Q-137 Are the CH_2OH and OCH_3 groups in this molecule equivalent, enantiotopic, or diastereotopic?

CH_2OH

H C OCH_3

CF_2CH_3

A-137 None of these, because the CH_2OH and OCH_3 groups do not have the same bonds. They are simply different groups.

Q-138 Are the two H's in the following molecule equivalent? Explain

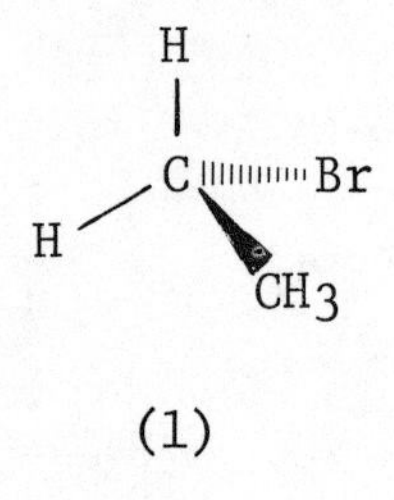

(1)

A-138 No, because their separate replacement by some achiral, new group, e.g., Cl, generates enantiomeric stereoisomers.

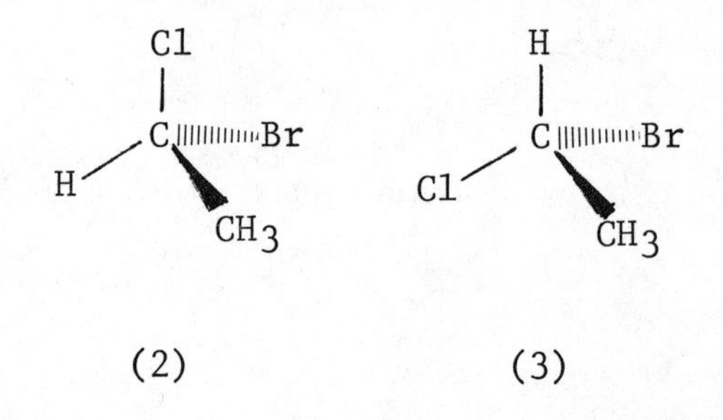

(2) (3)

Q-139 How should the two H's in (1) be classified?

A-139 They are enantiotopic, since replacement leads to enantiomers.

Q-140 Are any of the H's in the following compound equivalent?

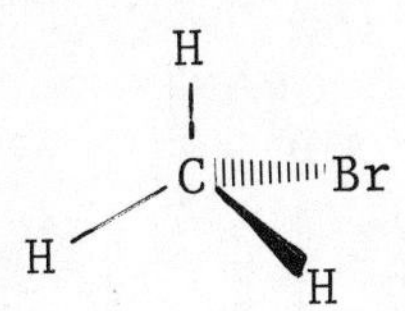

A-140 Yes, comparison of any pair of the three H's by the substitution criterion shows that all three are equivalent. Substitution of each H by Cl leads to the following compounds, none of which are isomers, because they are exactly the same.

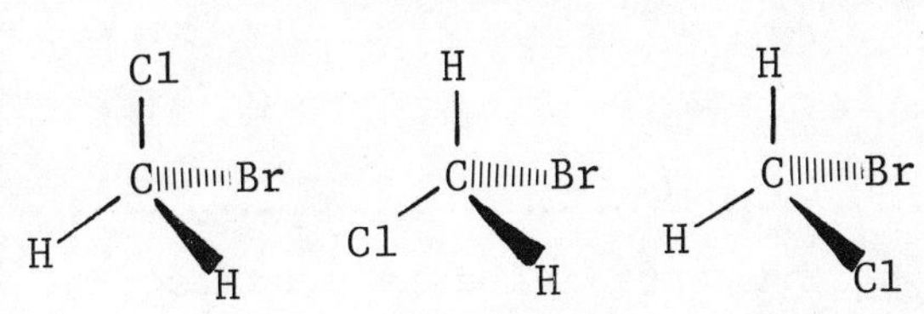

Q-141 Are the two CH_3 groups in the following compound enantiotopic?

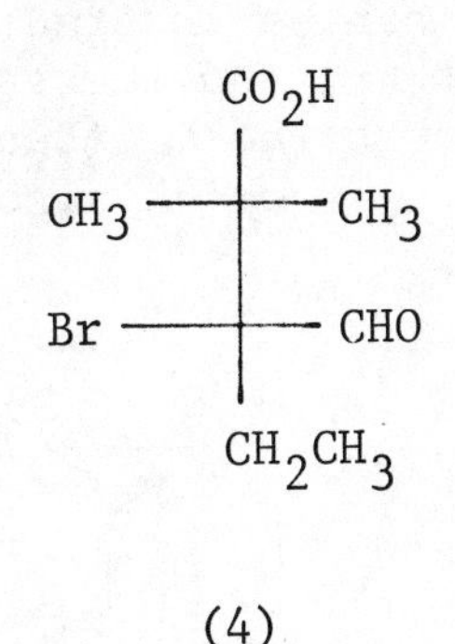

A-141 No, because the substitution test (here using H as the achiral substitution group) leads to a pair of diastereomers.

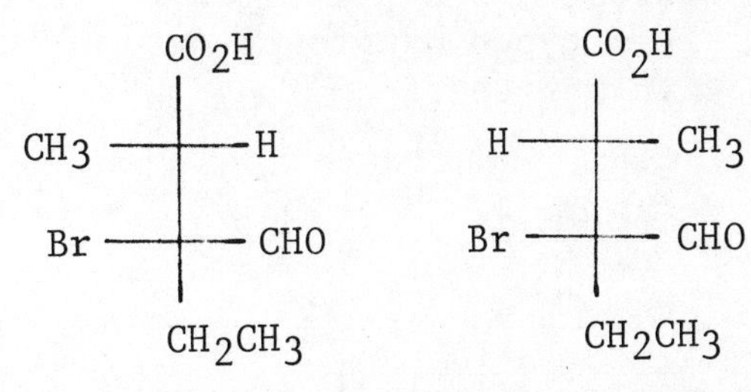

The two CH_3 groups therefore are diastereotopic.

Q-142 The hydrogens on the starred carbon in the diagram below are enantiotopic. How could they be made diastereotopic?

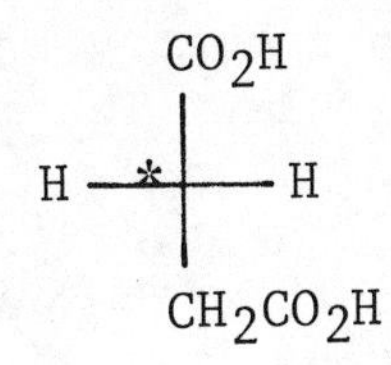

A-142 The hydrogens on the starred carbon can be made diastereotopic by replacing one of the hydrogens in the other CH_2 group with another atom.

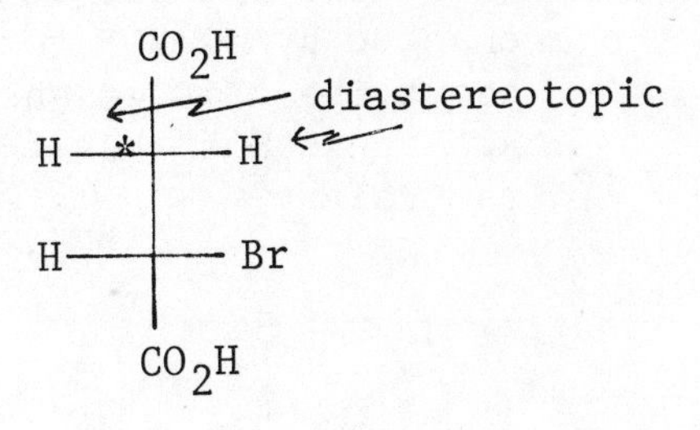

Q-143 How can methyl bromide, which has three equivalent hydrogens, be converted to one with enantiotopic hydrogens?

H

equivalent

H — Br

H

A-143 By replacing one of the hydrogens with another group, a compound with enantiotopic hydrogens is obtained.

H

enantiotopic

H · Br

NO_2

Q-144 Symmetry criteria are also available to distinguish equivalent, enantiotopic, and diastereotopic groups. For instance, equivalent groups always can be interchanged by a C_n symmetry operation ($n > 1$). Demonstrate this criterion for the two types of groups (atoms) in benzene.

Benzene

A-144 All of the carbon atoms are equivalent and all of the H atoms are equivalent in benzene. By using rotations of 180° (n = 2), 120° (n = 3), or 60° (n = 6), in either a clockwise or counterclockwise direction, any H can be interchanged with any other, and the same is true for any carbon. (Benzene has other C_n's which also could be used.)

C_n, n = 2, 3, 6

Q-145 These two H's are interchanged by a 180° rotation about the axis as shown. Are these H's therefore equivalent?

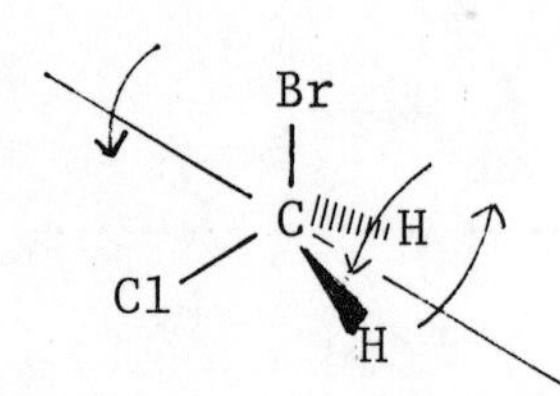

A-145 No, because the operation shown is not a symmetry operation for the molecule. The molecule has no C_n ($n>1$) axes. Consequently, the two H's are not equivalent. (They are enantiotopic.)

Q-146 Is there any S_n symmetry operation which will interconvert the two enantiotopic hydrogens in the above compound? If so, what is it?

A-146 Yes, an S_1 operation. Generally speaking, enantiotopic groups can be interchanged by an S_n symmetry operation. (Note: an S_1 operation is equivalent to a plane of symmetry perpendicular to the S_1 axis.)

Q-147 What symmetry operation will exchange the diastereotopic groups in this compound?

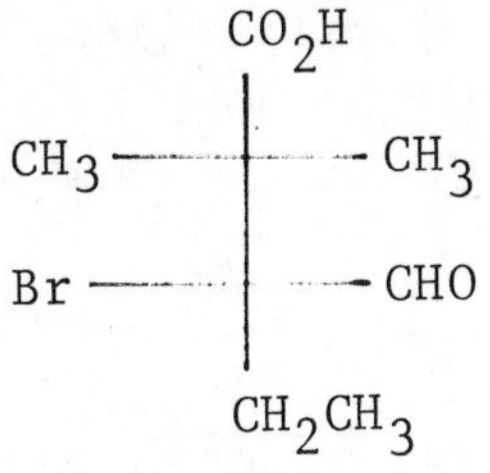

CO_2H

CH_3 — — CH_3

Br — — CHO

CH_2CH_3

A-147 None. Diastereotopic groups are not interchanged by any symmetry operation.

Q-148 What kinds of groups are the hydrogens in the following compound?

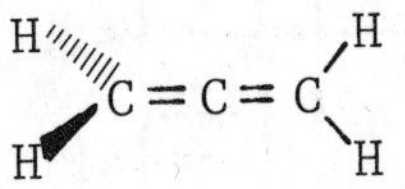

H, H C=C=C H, H

A-148 H_A and H_B are equivalent because they are interchanged by a C_2. H_A and H_c are enantiotopic because they are interchanged by an S_4.

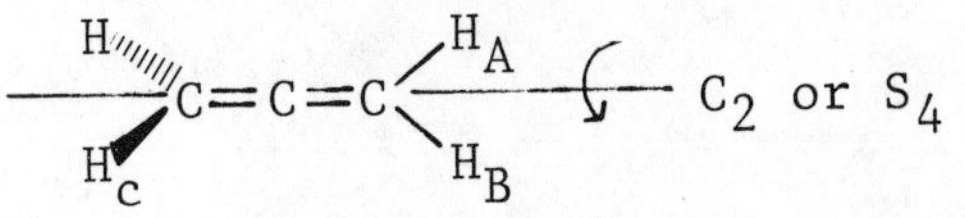

H, H_c C=C=C H_A, H_B — C_2 or S_4

D, L and Erythro, Threo:

S-18 Although the (R, S) specification of chirality is general and covers all classes of compounds, there are some other designations of chirality which are employed for special classes of compounds. For instance, the absolute chirality of carbohydrates and amino acids has been specified using the D, L scheme.

The carbohydrate (+)-glyceraldehyde was arbitrarily selected as a reference standard and designated D-(+)-glyceraldehyde. (-)-Glyceraldehyde was then designated L-(-)-glyceraldehyde.

CHO | CHO | CHO | CHO

H — C — OH ≡ H ◄ C ► OH   HO ◄ C ► H ≡ HO — C — H

CH_2OH | CH_2OH | CH_2OH | CH_2OH

D-(+)-Glyceraldehyde   L-(-)-Glyceraldehyde

The designation of compounds derived from glyceraldehyde as D or L is independent of the direction in which they rotate plane-polarized light.

In order to qualify as a derivative of D-glyceraldehyde, for instance, a compound must fit into the following generalized specification, where x is a hetero atom like O, N, or S and y is any atom.

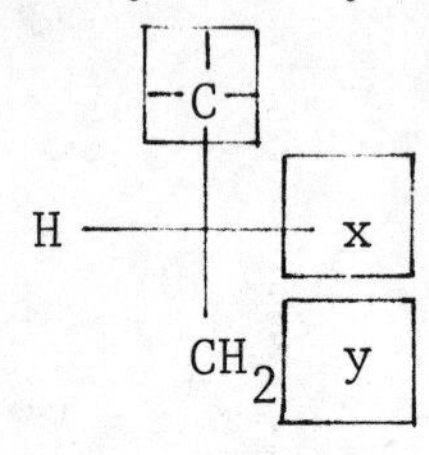

D-compound

Q-149 Draw D-(+)-lactic acid, which has the formula, $CH_3CHOHCO_2H$.

A-149

C = CO_2H

H — x = OH

CH_2 y = H

CO_2H

H — — OH

CH_3

D-(+)-lactic acid

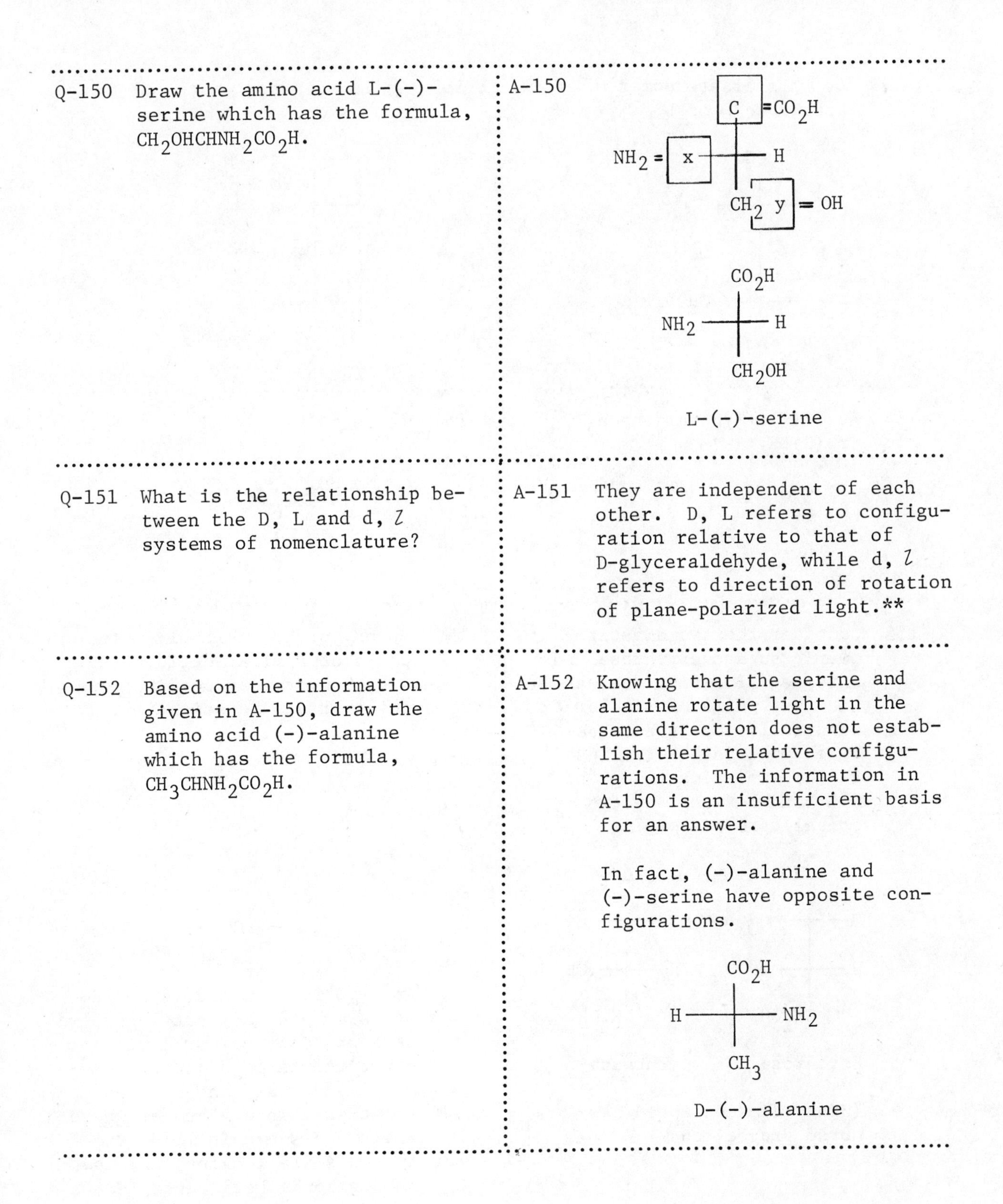

Q-150 Draw the amino acid L-(-)-serine which has the formula, $CH_2OHCHNH_2CO_2H$.

A-150

L-(-)-serine

Q-151 What is the relationship between the D, L and d, *l* systems of nomenclature?

A-151 They are independent of each other. D, L refers to configuration relative to that of D-glyceraldehyde, while d, *l* refers to direction of rotation of plane-polarized light.**

Q-152 Based on the information given in A-150, draw the amino acid (-)-alanine which has the formula, $CH_3CHNH_2CO_2H$.

A-152 Knowing that the serine and alanine rotate light in the same direction does not establish their relative configurations. The information in A-150 is an insufficient basis for an answer.

In fact, (-)-alanine and (-)-serine have opposite configurations.

D-(-)-alanine

**See A-78, p. 65.

Q-153 Is this diastereomer of arabinose D or L?

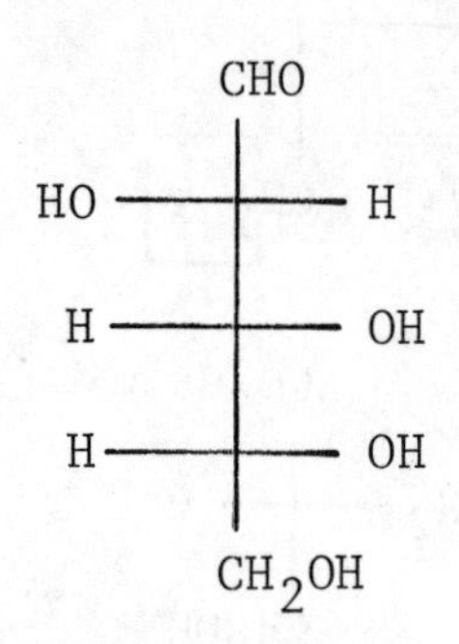

A-153 It is D, based on the CH_2OH end, where

Q-154 Is D-(+)-glyceraldehyde (R) or (S)?

A-154 It is (R).

Q-155 Consider the two diastereomeric sugars erythrose and threose. In which one can the most pairs of similar or identical substituents be lined up together (eclipsed) at the same time along the C-C bond between the two chiral carbon atoms?

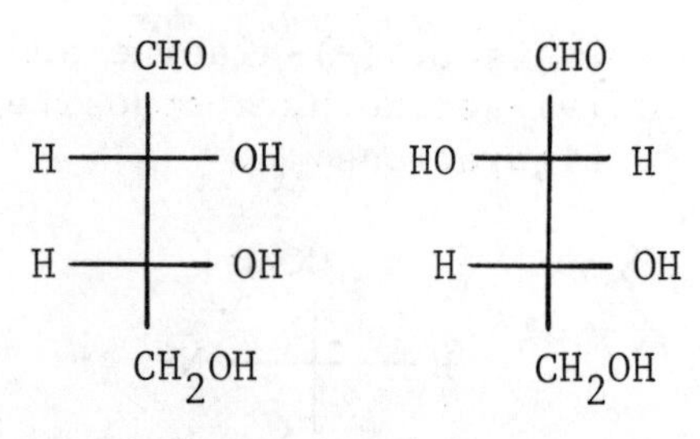

Erythrose Threose

(Hint: draw eclipsed sawhorse projections.)

A-155 In erythrose where all three pairs of similar or identical substituents can be lined up together at the same time.

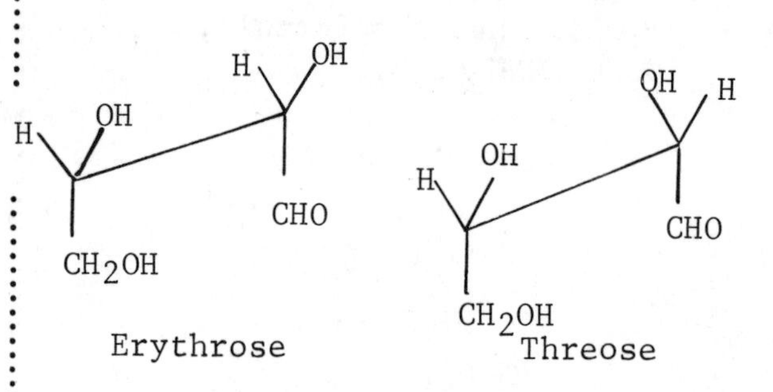

Erythrose Threose

An *erythro* compound is a diastereomer in which at least two sets of identical or similar substituents on adjacent asymmetric carbons can be lined up together. In erythrose, where the two H's are in line, the two OH's are in line, and the two carbon substituents (CH_2OH and CHO) are also in line, all at the same time. A *threo* compound is the same kind of diastereomer except only one pair can be lined up at one time, as in threose, where only the two carbon substituents (CHO and CH_2OH) are lined up in the conformation shown.

Q-156 Is this diastereomer *erythro* or *threo*?

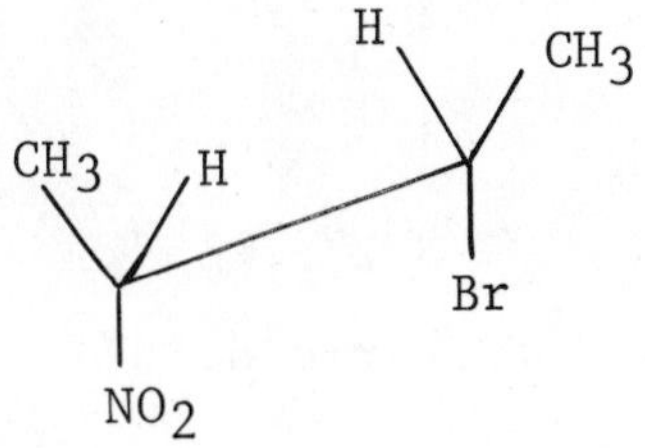

A-156 It is *threo*.

Q-157 Is this diastereomer *erythro* or *threo*?

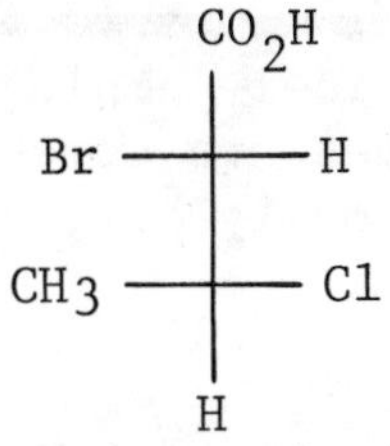

A-157 It is *erythro*.

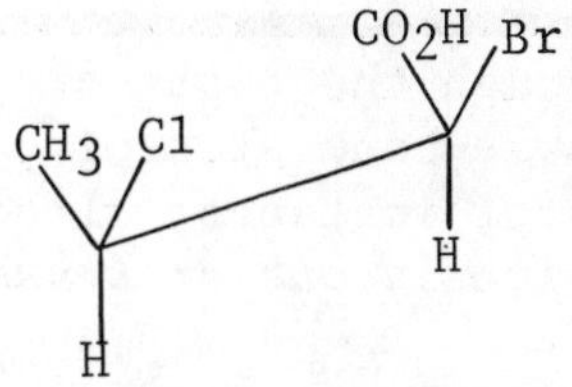

Q-158 Is this diastereomer *erythro* or *threo*?

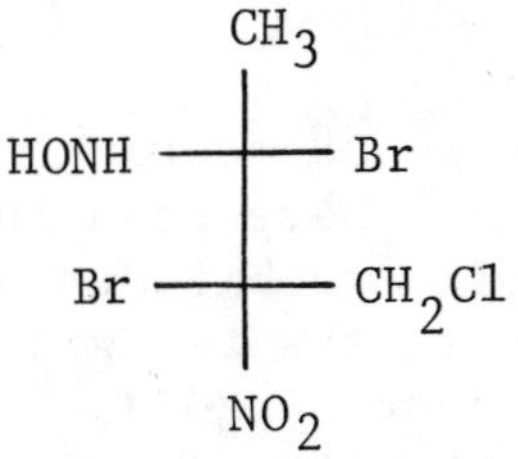

A-158 It is *erythro*.

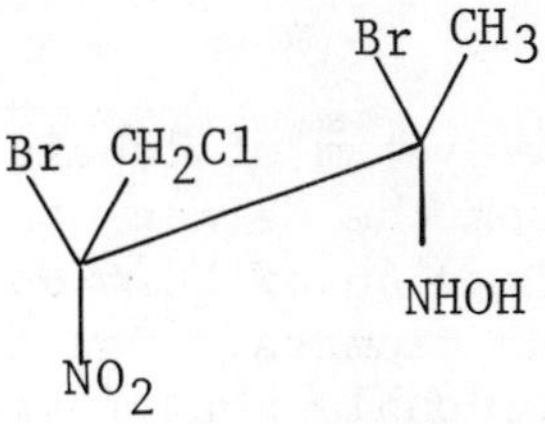

Q-159 Is this diastereomer *erythro* or *threo*?

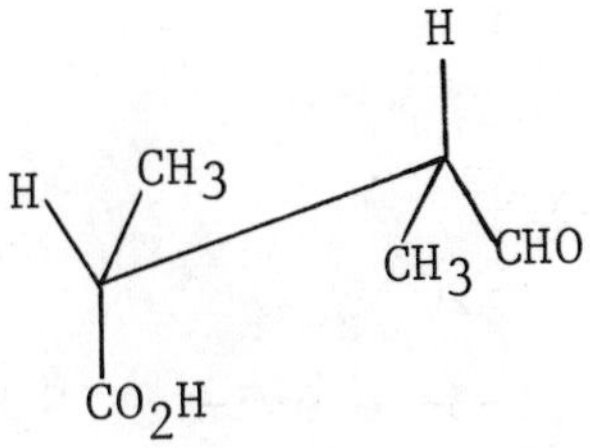

A-159 It is *threo*.

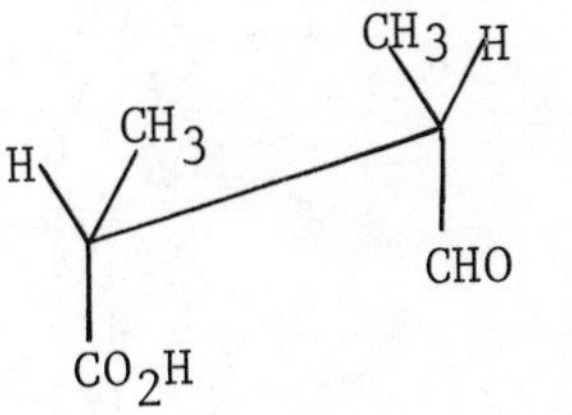

Q-160 Name and draw *meso*-tartaric acid in *erythro*, *threo* terminology.

$HO_2CCHOHCHOHCO_2H$

tartaric acid

A-160

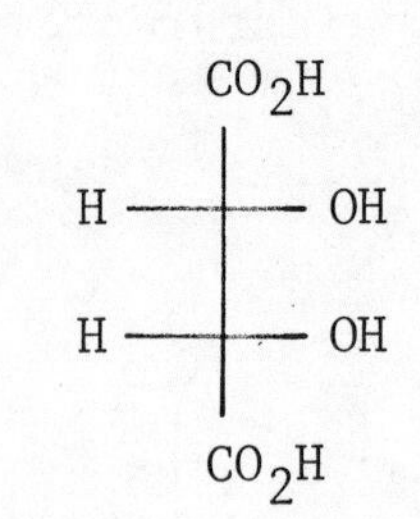

meso-tartaric acid

erythro-tartaric acid

The (E, Z) System:

S-19 Although the terms *cis/trans* and *syn/anti* are widely employed to describe double-bond stereoisomerism, the resulting names are sometimes ambiguous. For instance, although compound (1) is obviously *cis*, compound (2) is not obviously *cis* or *trans*.

```
CH3      CH3          CH3      Br
   \    /                \    /
    C=C                   C=C
   /    \                /    \
  H      H           C2H5      CO2H

    (1)                   (2)
```

A new system of labels (E) from the German *entgegen* meaning "opposite," and (Z) from the German *zusammen* meaning "together," have been introduced (as of Volume 66 of *Chemical Abstracts*) which permit unambiguous naming of double-bond isomers. One first assigns a priority to the two groups at one end of the double bond (*a* and *b*, according to the Cahn, Ingold, Prelog system) and then independently assigns a priority to the two groups at the other end of the double bond (*a* and *b* again). That configuration in which the two *a* groups are on the same side is (Z) and that configuration in which the two *a* groups are on opposite sides is (E). The priority assignments and names for compounds (1) and (2) are given below.

```
a CH3      CH3 a        b CH3      Br a
     \    /                  \    /
      C=C                     C=C
     /    \                  /    \
  b H      H b         a C2H5      CO2H b

 (Z)-2-butene         (E)-2-bromo-3-methyl-
                        2-pentenoic acid
```

Q-161 Is this compound (E) or (Z)?

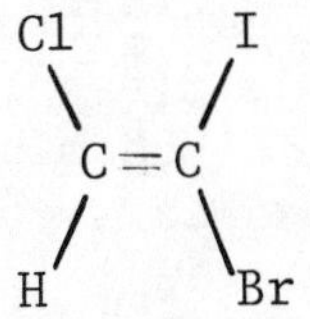

Cl, I on top; H, Br on bottom of C=C

A-161 It is (Z).

a, a on top; b, b on bottom of C=C

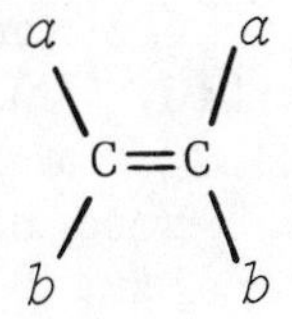

Q-162 Assign this compound (E) or (Z). (When no atom is present in a position, consider the position to have atomic number zero.)

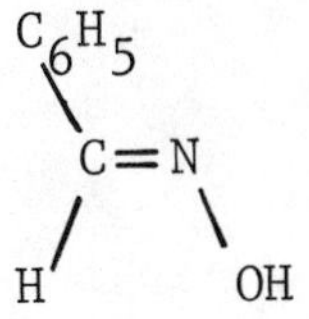

C_6H_5 and H on C; OH on N of C=N

A-162 It is (E).

a, b on C; a on N of C=N

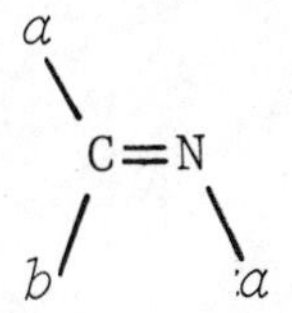

Q-163 Assign each of the double bonds in this 2,4,6-octatrienoic acid (E) or (Z).

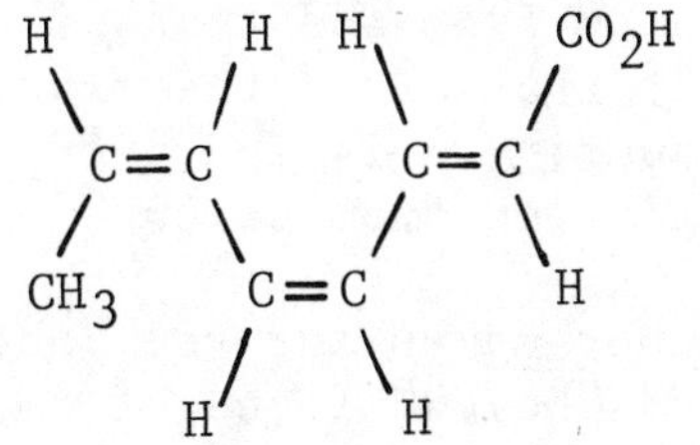

H, H, H, CO_2H; CH_3, H, H, H

A-163

b b b a; a a; a a; a b; Z E; b Z b

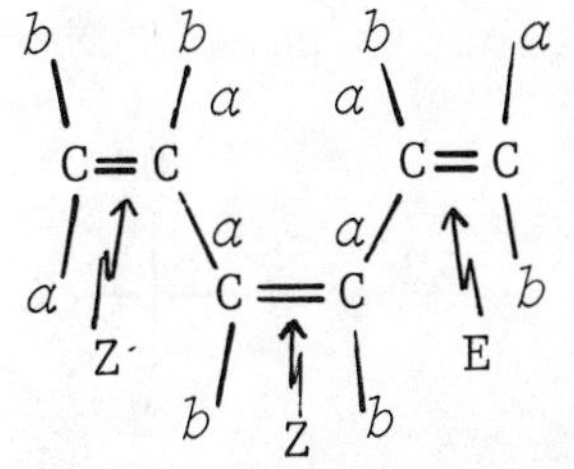

The compound is (2E, 4Z, 6Z)-octatrienoic acid.

Q-164 Assign this compound (E) or (Z).

CH_2SH; N; C; CO_2H

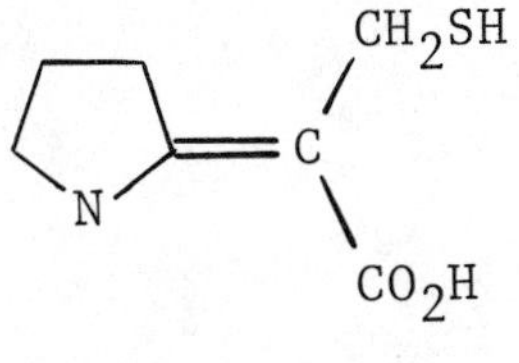

Atom	C	H	S	O	N
Atomic Number	6	1	16	8	7

A-164 It is (Z).

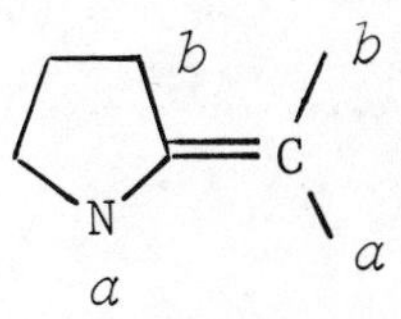

b, b; N a; C a

Note: C(HHS) < C(000)**

18 25

**See S-13, p. 79.

Faces of Double Bonds and *re*, *si*:

S-20 Just as there are equivalent, enantiotopic and diastereotopic groups, there also are equivalent, enantiotopic and diastereotopic faces of double bonds. This is based on whether addition of some achiral chemical reagent across the double bond from both sides results in equivalent, enantiomeric, or diastereomeric products.

Q-165 Draw the two faces of acetaldehyde, CH_3CHO.

A-165

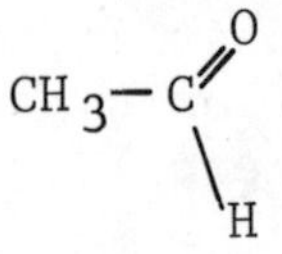

CH_3—C(=O)—H (1)   H—C(=O)—CH_3 (2)

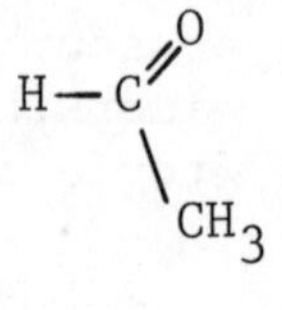

The two faces show the two sides of the carbonyl double bond.

Q-166 Given the reaction below, are the two faces of the carbonyl double bond in acetaldehyde equivalent?

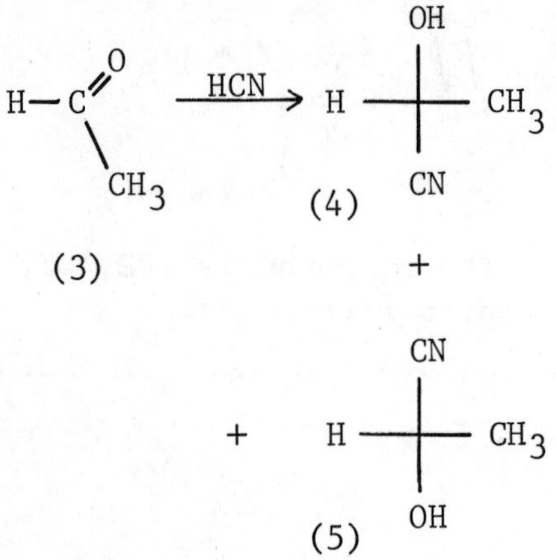

H—C(=O)—CH_3 (3) $\xrightarrow{HCN}$ H—C(OH)(CN)—CH_3 (4) + H—C(CN)(OH)—CH_3 (5)

A-166 No, they are enantiotopic, because compounds (4) and (5) are enantiomers. Compound (5) results from addition of HCN to (3) from above the plane of the paper; (4) is the product of addition from below.

Other names for the "above" and "below" faces of (3) are the "front" and "back" faces, respectively. Thus, (5) results from frontside addition of HCN.

Q-167 Given the following reaction, are the two faces of compound (6) enantiotopic?

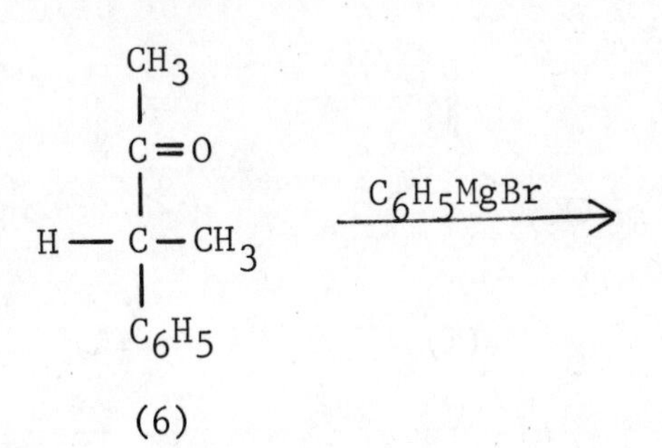

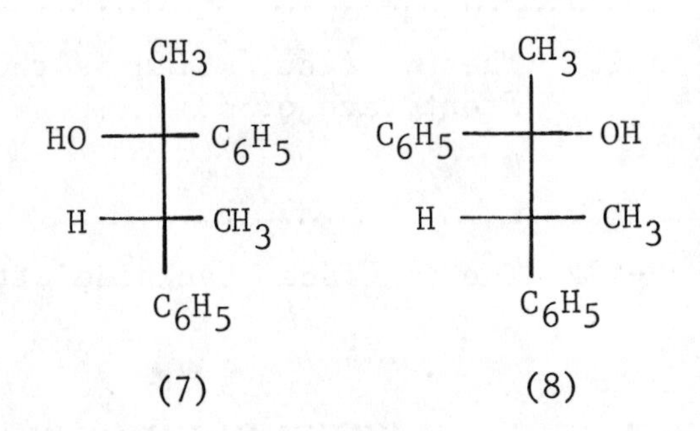

A-167 No, they are diastereotopic, because compounds (7) and (8) are diastereomers.

Q-168 Which product (7) or (8) resulted from addition of C_6H_5MgBr to the front face of (6)?

A-168 Product (8).

Q-169 How might the two faces of acetaldehyde (Q-166) be distinguished employing the Cahn, Ingold, Prelog sequence rule procedure?

A-169 One can simply use this procedure in two dimensions.

If the groups in order of decreasing priority, describe a clockwise arrangement, the face is called *re* (from Latin rectus). A counterclockwise arrangement defines a *si* face (*si* for Latin sinister). Note the correspondence to the R and S designations used in three dimensions.

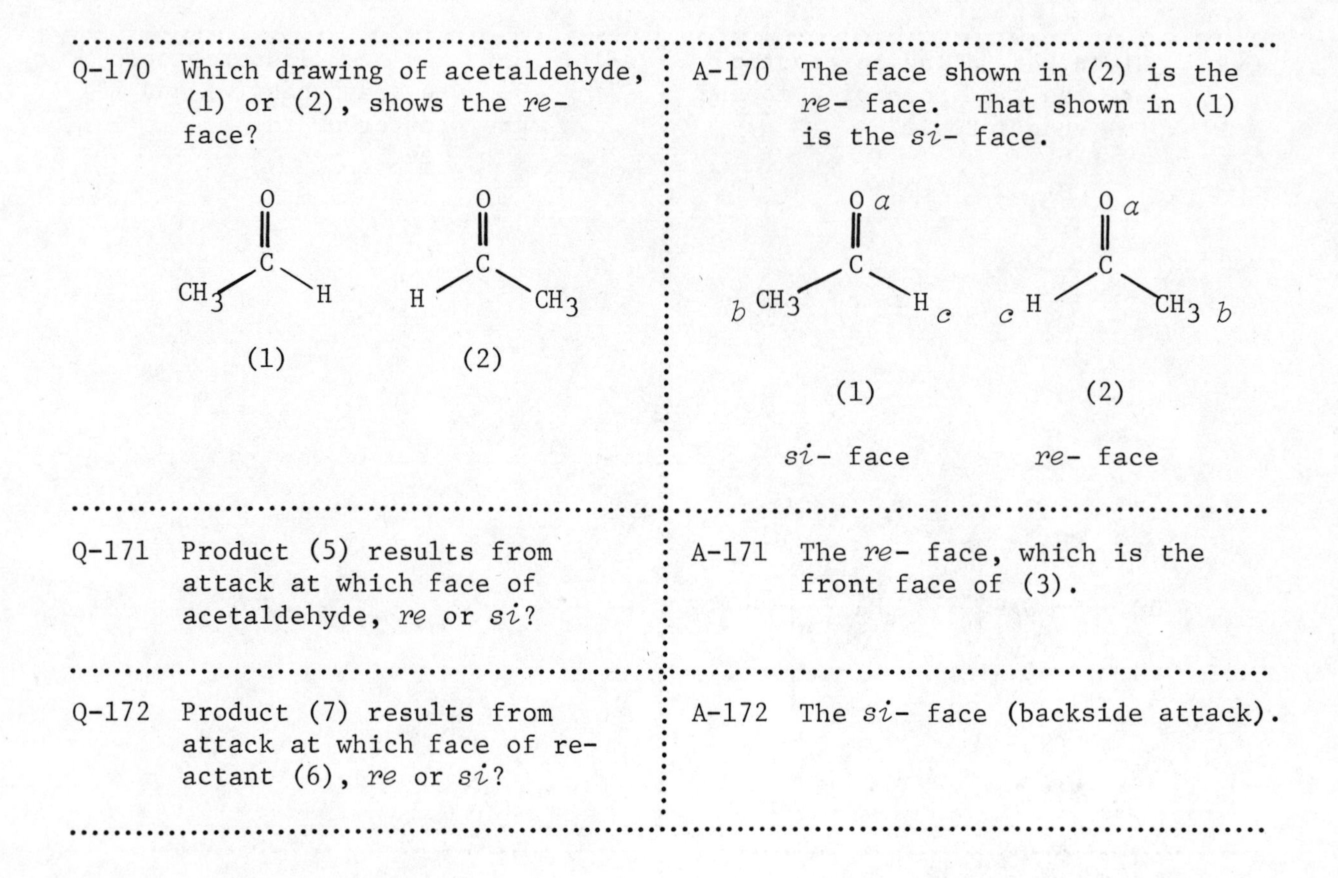

Q-170 Which drawing of acetaldehyde, (1) or (2), shows the *re*- face?

A-170 The face shown in (2) is the *re*- face. That shown in (1) is the *si*- face.

Q-171 Product (5) results from attack at which face of acetaldehyde, *re* or *si*?

A-171 The *re*- face, which is the front face of (3).

Q-172 Product (7) results from attack at which face of reactant (6), *re* or *si*?

A-172 The *si*- face (backside attack).

Section 4: Steric Aspects of Chemical Reactions

Racemic Modifications and Interconversion of Stereoisomers:

S-1 When equal amounts of enantiomeric molecules are present, the product is termed racemic, that is, a racemic modification, and is denoted by the symbols ($\pm$) or (dl).

$$(+)\text{-A} \quad + \quad (-)\text{-A} \quad = \quad (\pm)\text{-A}$$

$$(d)\text{-A} \quad + \quad (l)\text{-A} \quad = \quad (dl)\text{-A}$$

Enantiomers of compound A — Racemic modifications of compound A

Q-1 If a racemic modification is dissolved and the optical rotation of the solution measured, what would you expect to find for the optical rotation?

A-1 The optical rotation of the solution would be zero.

Q-2 Why does mixing equal amounts of two enantiomers, which are optically active, lead to an optically inactive mixture?

A-2 For every (+) molecule that rotates the light in one direction, there will also be a (-) molecule to rotate the light back an equal amount in the opposite direction. The net rotation is therefore zero.

Q-3 Show how you would name a racemic modification of alanine.

A-3 ($\pm$)-Alanine or (dl)-Alanine.

Q-4 Is the product of this reaction a racemic modification? Explain.

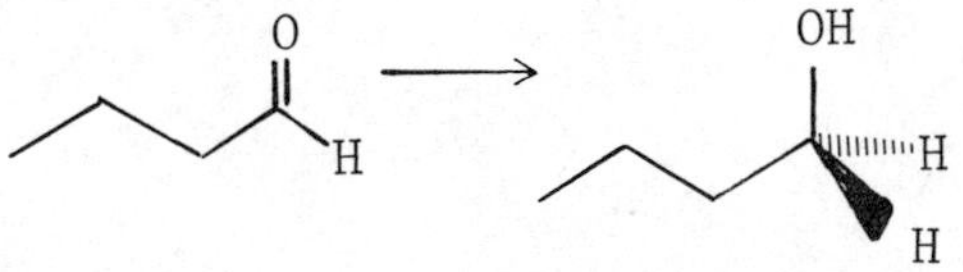

A-4 No, because no enantiomers are produced.

Q-5 Is the product of this reaction a racemic modification? Explain.

2-butanone $\xrightarrow{H_2}$ 2-butanol

A-5 Yes, in this case enantiomers are produced from symmetric reactants.

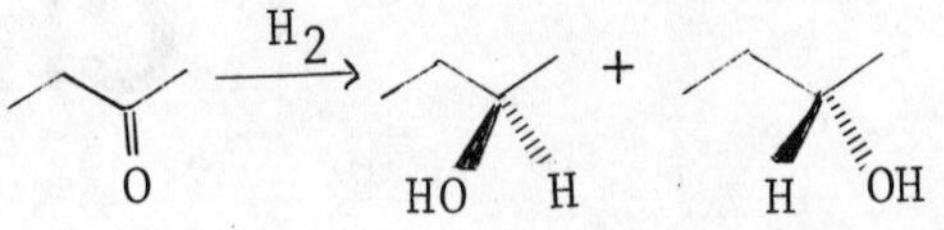

Since the 2-butanone is symmetric, the H_2 will attack it equally from both sides to give an equal mixture of the two enantiomeric alcohols.

Q-6 In a racemization reaction, an enantiomer is converted into a racemic modification. Give the other product in this racemization reaction.

flat

? +

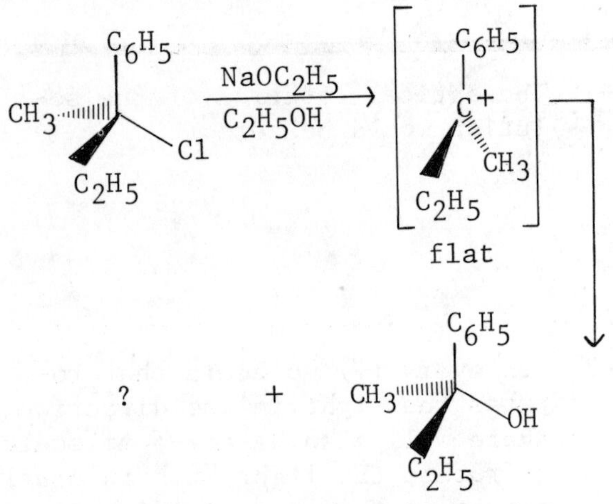

A-6 The other product is the second enantiomer.

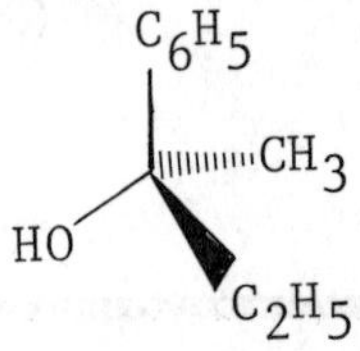

Q-7 Why do you suppose the above reaction proceeds with racemization?

A-7 The intermediate carbonium ion is symmetrical because it is flat, and therefore is attacked equally well from both sides by H_2O, leading to equal amounts of the two enantiomeric products. Sterically unrestricted carbonium ions are always flat.

Q-8 An unknown, optically active substance was dissolved in sodium hydroxide solution. The optical rotation decreased to zero, but the other spectral properties of the solution were unchanged. Explain this observation.

A-8 Racemization produced a racemic modification. Most likely this reaction proceeded through a carbanion intermediate. Carbanions, like carbonium ions, are flat. (See next question.)

Q-9 Show how this ethyl ester of (+)-mandelic acid can be racemized with sodium ethoxide in ethanol.

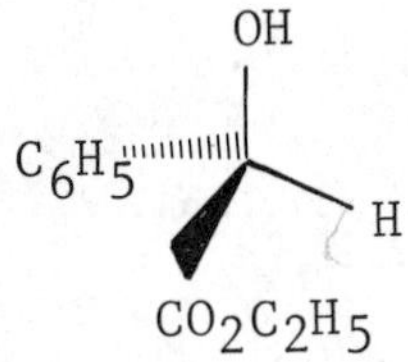

OH, C_6H_5, H, $CO_2C_2H_5$

A-9

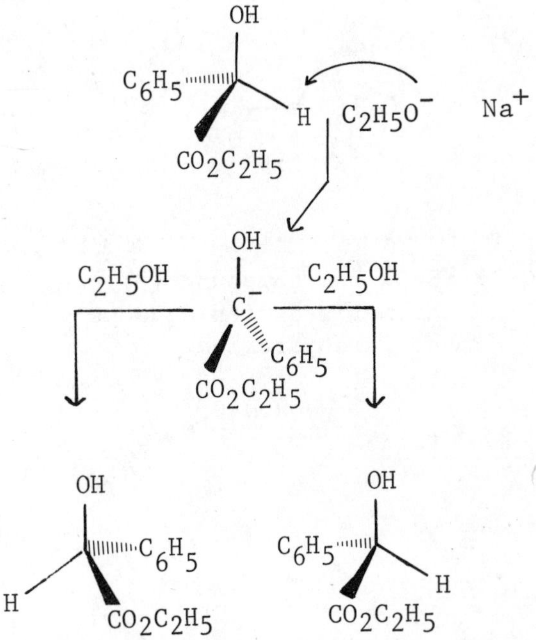

OH, C_6H_5, H, $CO_2C_2H_5$ + $C_2H_5O^-$ Na^+

↓

C_2H_5OH ← OH, C^-, C_6H_5, $CO_2C_2H_5$ → C_2H_5OH

OH, H, C_6H_5, $CO_2C_2H_5$ and OH, C_6H_5, H, $CO_2C_2H_5$

Q-10 Could the conversion of compound (1) into an equal mixture of compounds (1) and (2) qualify as a racemization process? Explain.

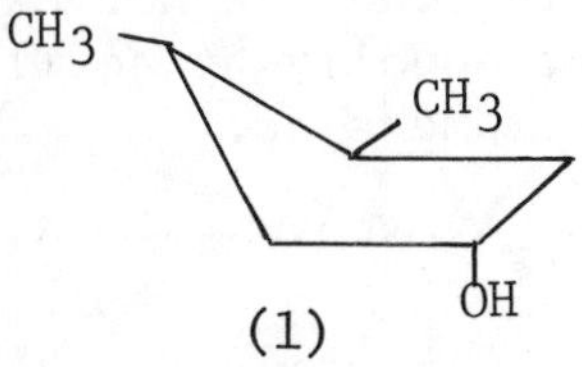

CH_3, CH_3, OH

(1)

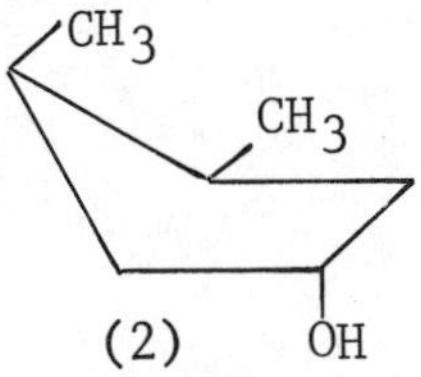

CH_3, CH_3, OH

(2)

A-10 No, because the two compounds are not enantiomers. They are diastereomers. Each has three chiral centers, and they differ in configuration at only one of them. These compounds are called epimers because epimers are diastereomers which contain two or more asymmetric atoms but differ in configuration at only one. The process described in Q-10 is an epimerization because it involves the interconversion of epimers.

Q-11 Which of these compounds are epimers?

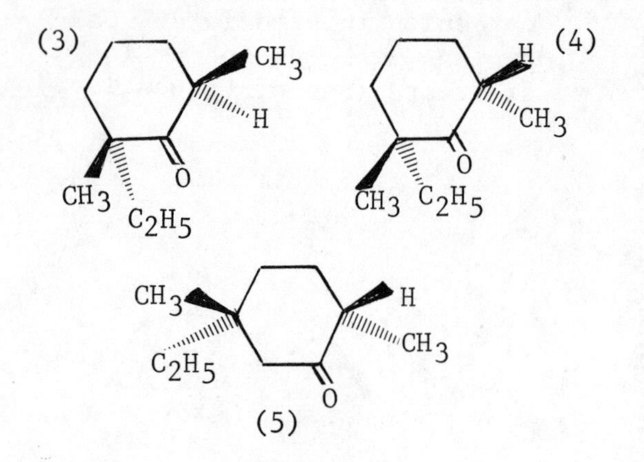

A-11 Compounds (3) and (4) are epimers. They contain two asymmetric carbons but differ in configuration at only one. Compound (5) is not even a diastereomer of either (3) or (4). Compound (5) is a position isomer of (3) and (4).

Q-12 Show how compound (3) can be epimerized with aqueous sodium hydroxide.

$$(3) \underset{}{\overset{NaOH, H_2O}{\rightleftharpoons}} (4)$$

A-12

CH_3, H, ^-OH, C_2H_5, CH_3, (3) $\rightleftharpoons$ CH_3, O^-, C_2H_5, CH_3 $\rightleftharpoons$ H, CH_3, C_2H_5, CH_3, (4)

Q-13 Why does not the above reaction result in racemization rather than epimerization?

A-13 The reaction involves diastereomers instead of enantiomers. Diastereomers differ in their chemical and physical properties and are not expected to occur equally at equilibrium as enantiomers do.

Q-14 Draw all the possible epimers of this compound.

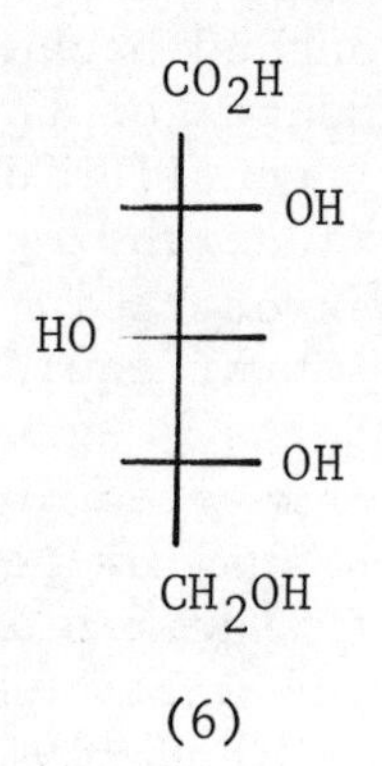

(6)

A-14 There are three epimers of compound (6).

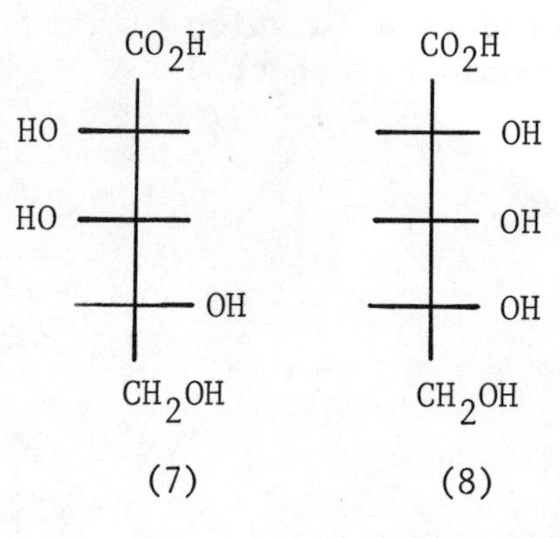

(7) (8)

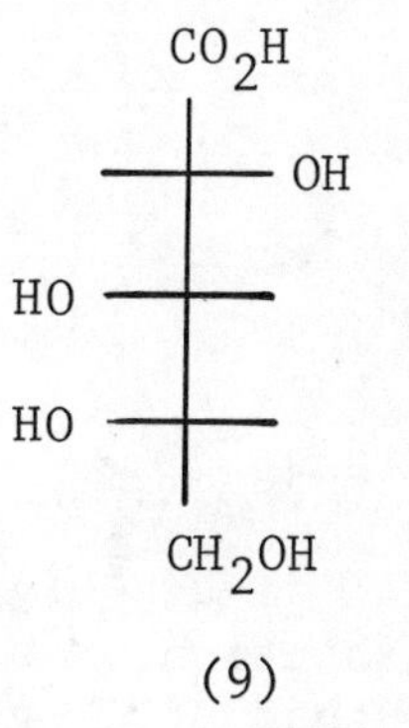

(9)

Q-15 The three forms of glucose shown below interconvert spontaneously in water. The open form is present in negligible concentration at equilibrium

α-Glucose (1)

$[\alpha]_D^{293} = +111°$

Open Form of Glucose (2)

β-Glucose (3)

$[\alpha]_D^{293} = +19.2°$

Is this interconversion a racemization, an epimerization, or neither?

A-15

Since the interconversion involves epimers, the interconversion is an epimerization. (Each ring form of glucose contains five chiral centers, only one of which undergoes a change in its configuration.)

...e is dissolved in ... optical rotation ... function of time, ... expected?

A-16 The rotation would change with time until eventually a stable value is reached because the relative amounts of the epimers shift until equilibrium is reached. Since the open form of glucose is present in negligible concentration at equilibrium, and some of the α-glucose will have been converted to β-glucose which has a smaller positive rotation, then the final equilibrium rotation will be somewhere between +111° and +19.2°.

This process in general is called a mutarotation because mutarotation involves a change in optical rotation of a solution as a function of time. Usually an epimerization is involved.

Q-17 Which will give a higher final* optical rotation, a solution of α-glucose or β-glucose?

* After mutarotation.

A-17 Neither. The same proportions of α- and β-forms will result whether one started with α-glucose or with β-glucose because a single equilibrium will be reached. Therefore, the same optical rotation will be obtained in either case.

Q-18 The actual rotation of an aqueous glucose solution at equilibrium is $^{293}_{D} = +52.5^{\circ}$. From this value calculate the proportions of α- and β-glucose present at equilibrium. The open form is present in negligible concentration so it can be ignored.

(Hint: assume that the contributions of the α and β epimers to the observed rotation are proportional to their concentrations.)

A-18 Let x = fraction of glucose in the α form at equilibrium.

Then 1 - x = fraction of glucose in the β form at equilibrium.

$$(+111^{\circ})(x) + (+19.2^{\circ})(1-x) = +52.5^{\circ}$$

$$111x + 19.2 - 19.2x = 52.5$$

$$91.8x = 33.3$$

$$x = .362 \quad (36.2\%\ \alpha)$$

$$1 - x = .638 \quad (63.8\%\ \beta)$$

Resolution of Racemic Modifications:

S-2 To resolve a racemic modification means to separate the racemic modification into enantiomers. Since enantiomers can be *distinguished* only by interaction with chiral agents, enantiomers can be *separated* only by interaction with chiral agents. In the simplest example, an enantiomer may interact preferentially with itself and separate by crystallization. However, resolution of racemic modifications is usually carried out by treating the enantiomers with chiral agents such as chiral chemical modifiers, chiral complexing or adsorbing agents, and chiral inclusion compounds.

Q-19 Imagine the recrystallization of a racemic compound from water by slow evaporation. After a time two different kinds of crystals formed whose faces were mirror images of each other and lacked all symmetry. What property might be different for the two types of crystals and how could you determine this? Provide experimental details.

A-19 Carefully separate the two types of crystals using a tweezers. Dissolve an equal weight of each type in an equal quantity of water and then measure the optical rotation of each solution. The two types of crystals might be enantiomers and have equal but opposite rotations.

Q-20 In fact this was exactly the experience of Pasteur in 1848 after he observed mirror image crystals forming from a solution of racemic sodium ammonium tartrate. He observed that the optical rotation of the two solutions prepared from the separated crystals were equal but opposite in sign. Had Pasteur actually resolved the racemic salt?

A-20 Yes, he had. Each enantiomer had crystallized by itself. The entire crystalline product, prior to separation with a tweezers, was a racemic mixture. The resolution was therefore by manual separation. The two separated piles of crystals contained pure enantiomers.

Q-21 Subsequent research has shown that an aqueous solution of sodium ammonium tartrate yields mirror image crystals only below 301 K. Above 301 K, only symmetrical crystals are obtained. Offer an explanation.

A-21 The crystals which result above 301 K contain equal amounts of each enantiomer. Apparently the interaction between crystallizing molecules of the same configuration is weakened sufficiently with increasing temperature that above 301 K the interaction between enantiomeric molecules is stronger. Crystals of racemic composition result.

Q-22 Actually the method of mechanical separation is seldom useful. Even though crystals of pure enantiomers are always mirror images of each other, the distinctive crystal faces are usually so small and poorly developed as to be useless from a practical standpoint. Can you suggest a more useful way of resolving a racemic mixture by crystallization?

(Hint: assume you have a sample of one of the pure enantiomers.)

A-22 Yes, by inducing only one enantiomer to crystallize. This can sometimes be accomplished by adding a small seed crystal of one enantiomer to a saturated solution of the mixture. Resolution takes place as long as only one of the enantiomers crystallizes. The other enantiomer does not crystallize simply because nothing has induced its crystallization. Typical "natural" inducers of crystallization are dust particles and lattice sites on glass containers.

Sometimes one enantiomer can be induced to crystallize by seeding with a different compound of similar crystal structure.

Some Common Resolving Agents:

S-3

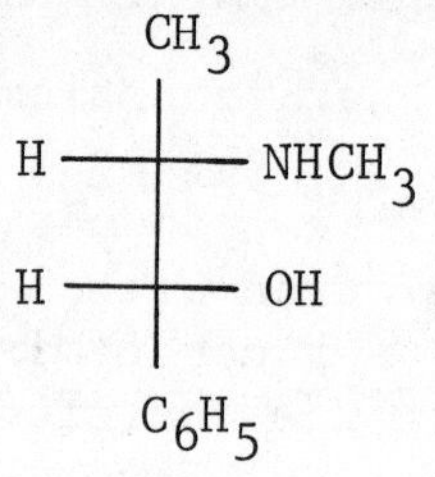

(1) (-)-Ephedrine

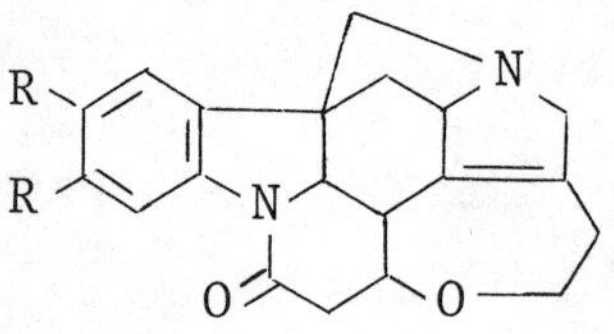

(2) R=H Strychnine
(3) R=CH_3O Brucine

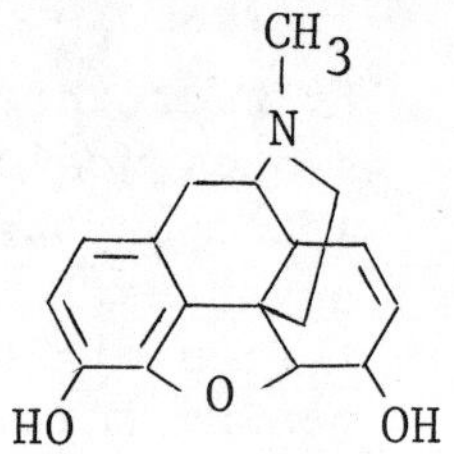

(4) Morphine

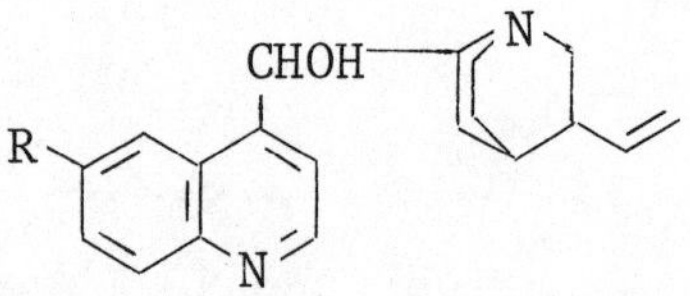

(5) R=H Cinchonine
(6) R=CH_3O Quinine

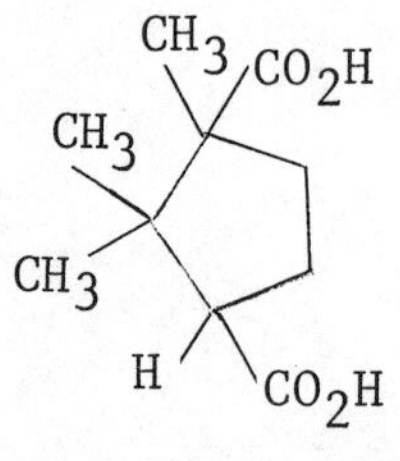

(7) Camphoric Acid

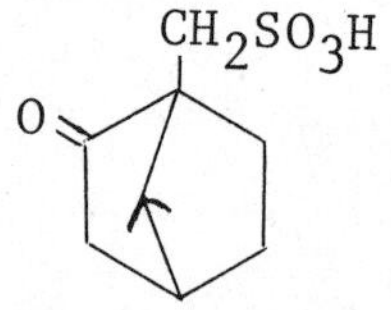

(8) Camphor-10-Sulfonic Acid

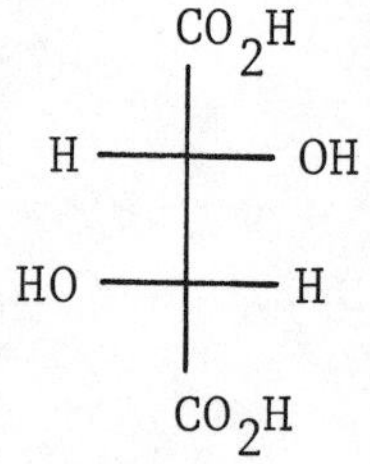

(9) Tartaric Acid

Q-23 Imagine that you have a racemic mixture of compound A, (±)-A, which you want to resolve. You know that fractional crystallization is a good separation procedure but rarely works for enantiomers because they have identical chemical and physical properties. However, you have a sample of one of the enantiomers of compound B, say (-)-B, and you know that A reacts with B to form a crystalline salt, AB. What would you obtain if you reacted (±)-A with (-)-B?

A-23 You would obtain a mixture of these two salts.

(+)-A(-)-B

+

(-)-A(-)-B

Q-24 Why might it be easier to separate these two salts by fractional recrystallization than it was to resolve the original racemic mixture by fractional recrystallization?

(Clue: is this new mixture also a racemic mixture?)

A-24 The two salts are diastereomers. Their mixture is therefore not a racemic mixture. Since diastereomers have different chemical and physical properties, they are usually easier to separate than enantiomers.

Q-25 If one ultimately wants to obtain pure (+)-A and pure (-)-A from a racemic mixture of (±)-A, why is it particularly helpful when (+)-A(-)-B and (-)-A(-)-B are salts?

A-25 Salts are usually easy to form, crystallize, and to return to starting materials.

Q-26 What product is obtained from reaction of (±)-glyceric acid with (+)- α -phenethylamine?

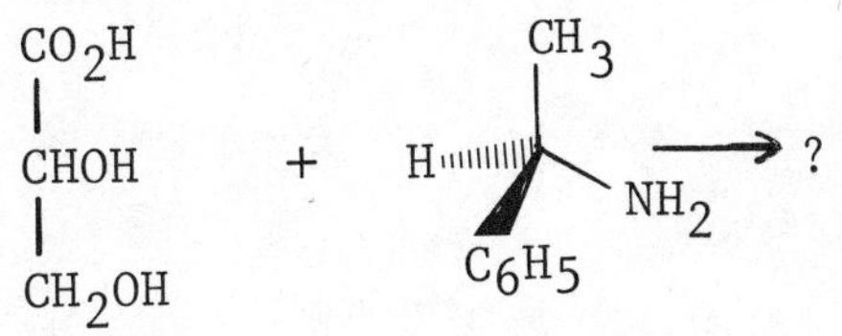

CO_2H
|
CHOH
|
CH_2OH

+

CH_3
H
NH_2
C_6H_5

→ ?

Racemic Modification

(±)-glyceric acid
or
(±)-GCO_2H

Resolving Agent

(+)- α -phenethylamine
or
(+)-PNH_2

A-26 The product is a mixture of the two diastereomeric salts shown below.

(+)-GCO_2^- · (+)-PNH_3^+

+

(-)-GCO_2^- · (+)-PNH_3^+

Q-27 What is the next step if one is trying to resolve (±)-glyceric acid?

A-27 Separate the diastereomeric salts by fractional crystallization.

Q-28 Why would this procedure allow one to recover $(+)\text{-}GCO_2H$, that is, to separate $(+)\text{-}GCO_2H$ from the resolving agent?

$$(+)\text{-}GCO_2^- \cdot (+)\text{-}PNH_3^+$$

$\downarrow$ 10% aqueous HCl

$$(+)\text{-}GCO_2H + (+)\text{-}PNH_3^+Cl^-$$

$\downarrow$ Extract with an organic solvent like ethyl acetate.

A-28 The uncharged $(+)\text{-}GCO_2H$ partitions into the ethyl acetate while the charged $(+)\text{-}PNH_3^+Cl^-$ stays in the aqueous phase.

Q-29 Which of the common resolving agents in S-3 could be used to resolve $(\pm)\text{-}RCO_2H$ and which could be used to resolve $(\pm)\text{-}RNH_2$?

A-29

Racemic Modification	Potential Resolving Agents
$(\pm)\text{-}RCO_2H$	(1) (2) (3) (4) (5) (6)
$(\pm)\text{-}RNH_2$	(7) (8) (9)

In each case, ammonium carboxylate salts are formed.

Q-30 Why do you suppose the alkaloids (1) (2) (3) (4) (5) and (6) have been employed as resolving agents?

A-30 These natural products occur in nature in optically pure forms, are easy to isolate in reasonable quantities, and have been found to form salts which crystallize easily, partly due to their rigid bulk.

Q-31 A chemist who faced the task of resolving (±)-2-butanol went to the stockroom and came back with brucine and phthalic anhydride. Brucine does not form a salt with (±)-2-butanol. What did the chemist have in mind?

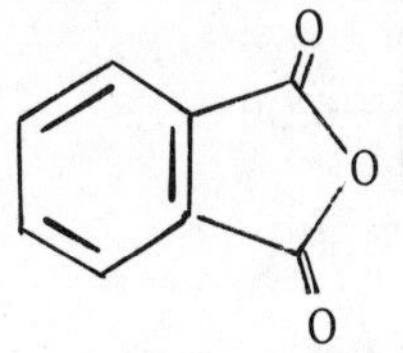

phthalic anhydride

A-31 Phthalic anhydride is first used to place a carboxylic acid "handle" on the alcohol.

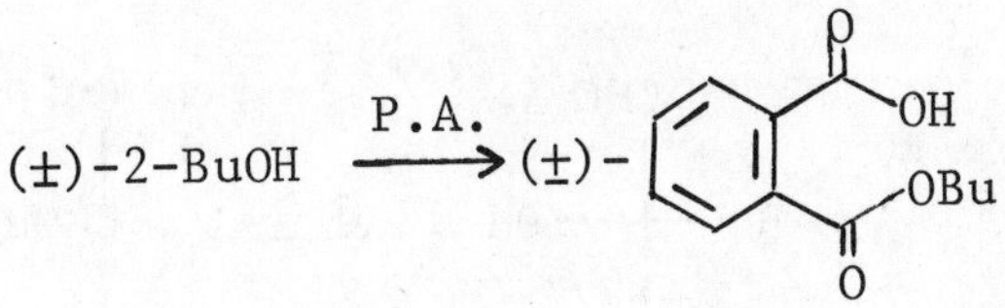

(±)-2-BuOH —P.A.→ (±)- [phthalate half-ester: C(=O)OH, C(=O)OBu]

Brucine can then be used to form a diastereomeric ammonium carboxylate salt, followed by the usual fractional crystallization and recovery of separated enantiomers.

Q-32 When chromatographed on filter paper, racemic 2,3-dihydroxyphenylalanine (10), gave two spots. Does this indicate that resolution has been effected in the absence of a chiral agent? Explain.

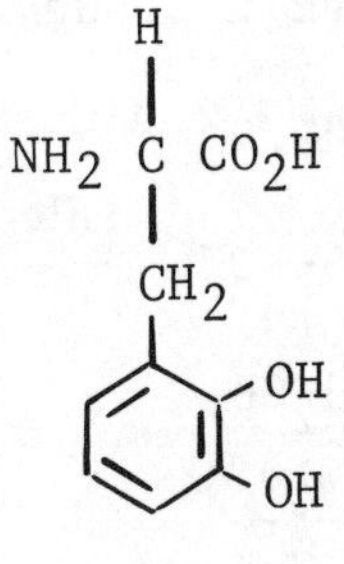

H–C(NH$_2$)(CO$_2$H)–CH$_2$–C$_6$H$_3$(OH)$_2$

(10)

A-32 No, because paper is cellulose, a natural product which is optically active.

Relative Configurations I:

S-4 To know the relative configuration of a compound is to know whether the compound has the same or opposite configuration relative to a second compound. Generally the relative configurations of compounds are established through chemical reactions.

Q-33 Which enantiomer of (2) would result from the reaction of HgO with (1)? The reaction is given below

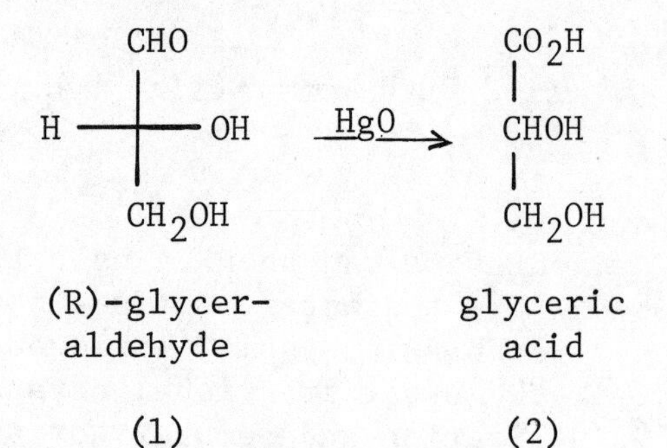

A-33 The (R)-enantiomer of (2) would result. The reaction does not break any of the bonds to the asymmetric carbon, and the (R)-enantiomer of (2) has the same arrangement of corresponding groups as does (R)- (1).

CO_2H

H — OH

CH_2OH

(R)- (2)

Q-34 Which enantiomer of (2) would result from the (+) enantiomer of (1)?

A-34 One cannot answer this question unless one knows whether (+)- (1) is (R)- (1) or (S)- (1).

Q-35 It is known that (+)- (1) is (R)- (1). Now determine whether (+)- (1) would afford (+)- (2) or (-)- (2).

A-35 This question still cannot be answered since it is not known whether (R)- (2) is (+)- (2) or (-)- (2). In fact, (R)- (2) is (-)- (2). Thus,

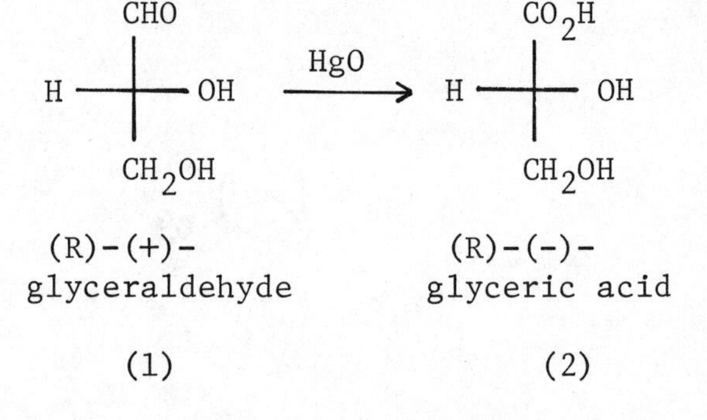

Q-36 A student has a sample of (-)- (4) and (R)-(-)- (3). Suggest a way for him to assign the relative configuration of (-)- (4). (Hint: Use the table of reagents in the Appendix to find a means to interconvert these two compounds.)

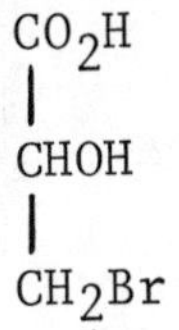

$$\begin{array}{c} CO_2H \\ | \\ CHOH \\ | \\ CH_2Br \end{array} \qquad \begin{array}{c} CO_2H \\ | \\ CHOH \\ | \\ CH_3 \end{array}$$

(3) 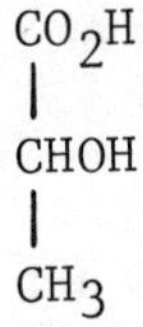 (4)

A-36

$$\begin{array}{c} CO_2H \\ H - \!\!\!\! + \!\!\!\! - OH \\ CH_2Br \end{array} \xrightarrow[C_2H_5OH]{Na \cdot Hg} \begin{array}{c} CO_2H \\ H - \!\!\!\! + \!\!\!\! - OH \\ CH_3 \end{array}$$

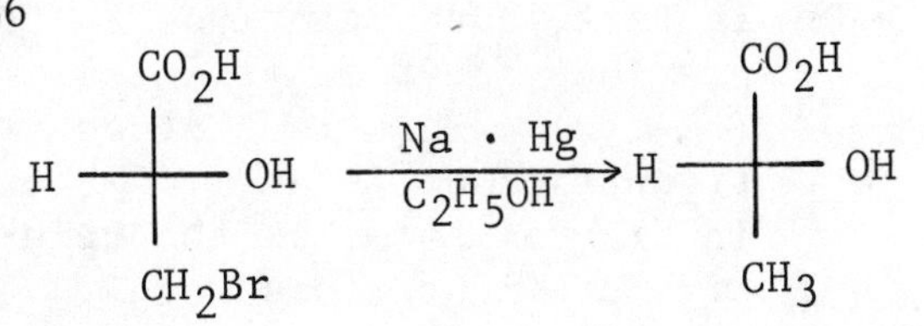

(R)-(-)-(3) (S)-(4)

Since this reaction does not break any bonds to the asymmetric carbon, the reaction can be used to correlate the configurations of the two compounds. As a final step the student would measure the optical rotation of (S)- (4) to determine if (S)- (4) is (-)- (4) or (+)- (4).

Q-37 Correlate the configurations of compound (1) and the (+) enantiomer of (5). (Hint: Convert both (1) and (5) to a common third compound.)

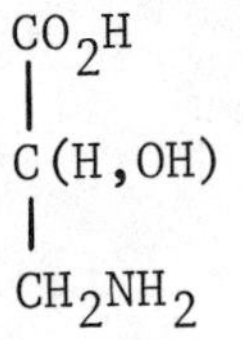

$$\begin{array}{c} CO_2H \\ | \\ C(H,OH) \\ | \\ CH_2NH_2 \end{array} \qquad \begin{array}{c} CHO \\ H - \!\!\!\! + \!\!\!\! - OH \\ CH_2OH \end{array}$$

Isoserine (R)-(+)-Glyceraldehyde

(5) (1)

A-37 First, both compounds are converted to glyceric acid (2), using reactions that do not break any of the bonds to the asymmetric carbon.

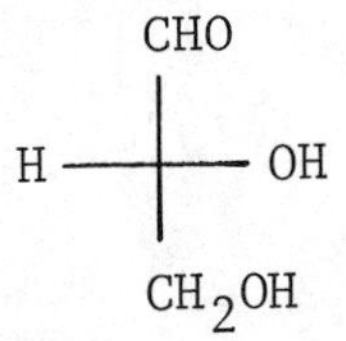

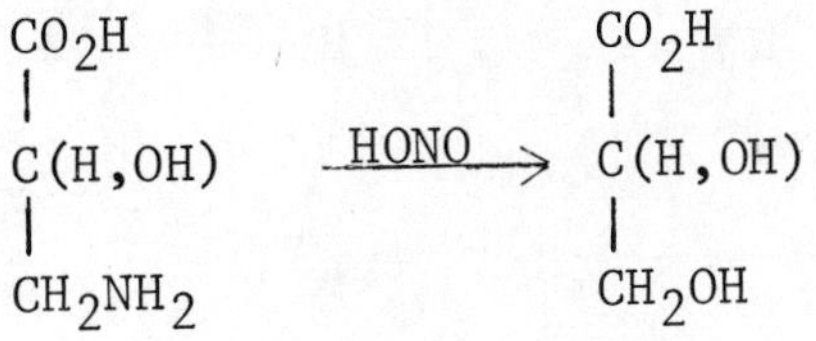

$$\begin{array}{c} CO_2H \\ | \\ C(H,OH) \\ | \\ CH_2NH_2 \end{array} \xrightarrow{HONO} \begin{array}{c} CO_2H \\ | \\ C(H,OH) \\ | \\ CH_2OH \end{array}$$

(5) (2)

$$\begin{array}{c} CHO \\ H - \!\!\!\! + \!\!\!\! - OH \\ CH_2OH \end{array} \xrightarrow{HgO} \begin{array}{c} CO_2H \\ H - \!\!\!\! + \!\!\!\! - OH \\ CH_2OH \end{array}$$

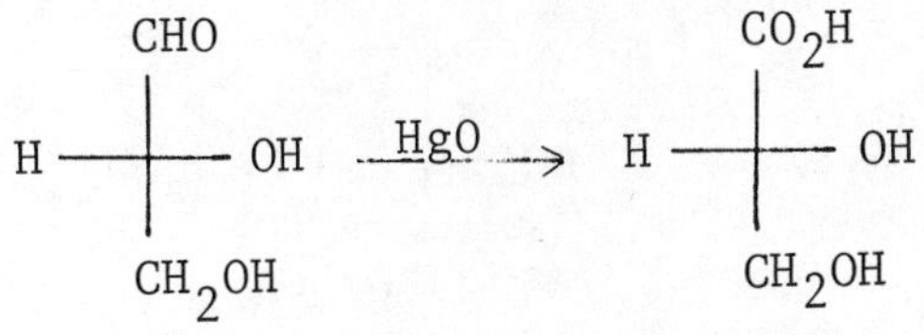

(1) (2)

Then the optical rotation of the product of each reaction is measured. If both products have the same sign, (1) and (+)- (5) have the same configuration.

Q-38 In fact, both reactions in the previous question produce (-)- (2). What is the configuration of (+)- (5). Show in diagram form how the relative configurations were established.

A-38

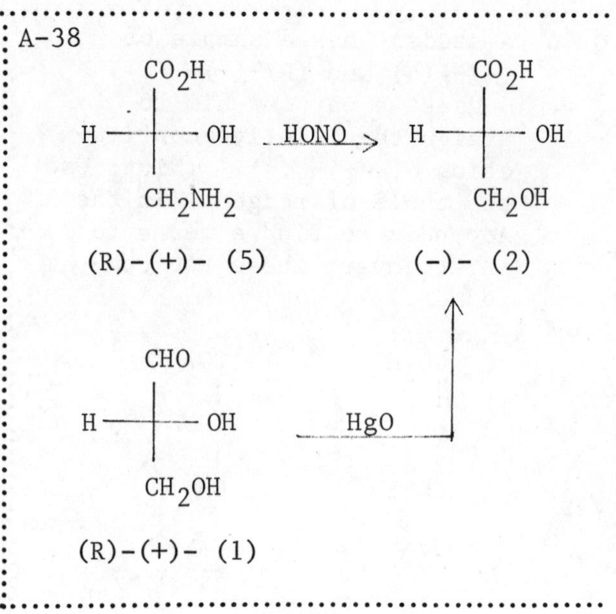

Q-39 Interconvert (+)-tartaric acid (6) and (+)-malic acid (7). Hint: Assume you can form and isolate compound (8).

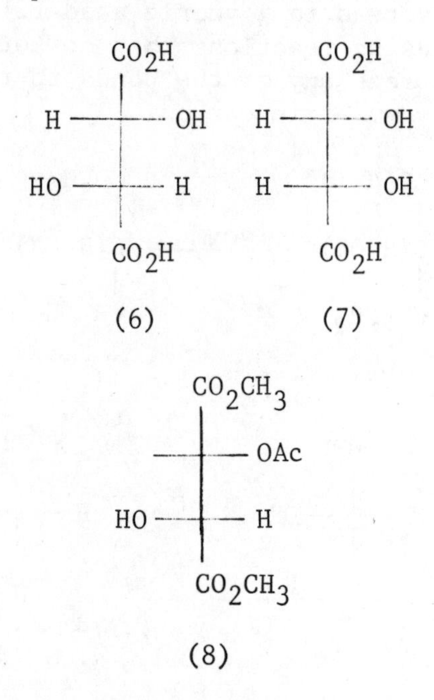

A-39

Step 1. (6) $\xrightarrow[\text{2. 1/2 equiv. } Ac_2O]{\text{1. HCl, } CH_3OH}$

CO_2CH_3 / H — OH / AcO — H / CO_2CH_3 (9) + CO_2CH_3 / H — OAc / HO — H / CO_2CH_3 (8)

Step 2. Separate the diastereomers (8) and (9).

Step 3.

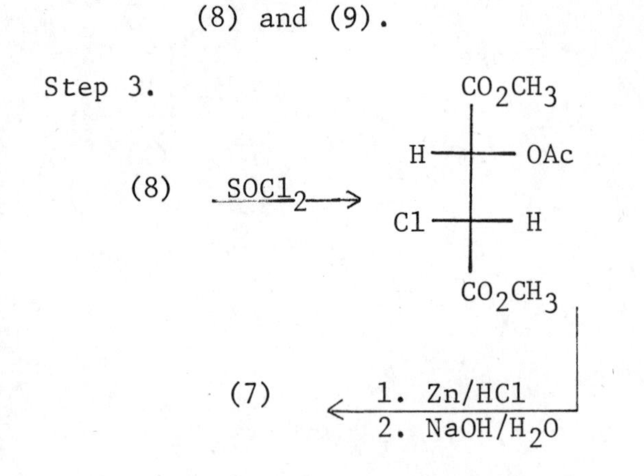

Q-40 Interconvert (+)-malic acid (7) and (+)-isoserine (5).

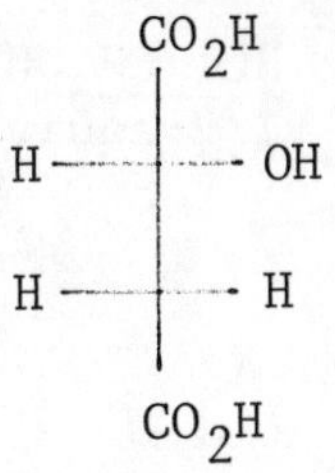

$$\begin{array}{c} CO_2H \\ H - C - OH \\ H - C - H \\ CO_2H \end{array} \qquad \begin{array}{c} CO_2H \\ H - C - OH \\ H - C - H \\ NH_2 \end{array}$$

(7) 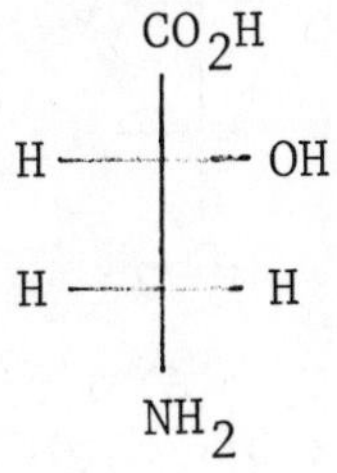 (5)

A-40

Step 1. (7) $\xrightarrow{\text{1/2 equiv. } HCl/CH_3OH}$

$$\begin{array}{c} CO_2H \\ H - C - OH \\ CH_2 \\ CO_2CH_3 \end{array} + \begin{array}{c} CO_2CH_3 \\ H - C - OH \\ CH_2 \\ CO_2H \end{array}$$

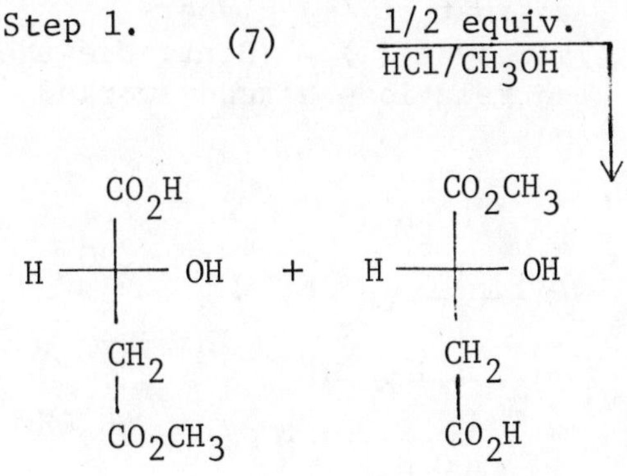

(10) (11)

Step 2. Separate (10) and (11)

Step. 3

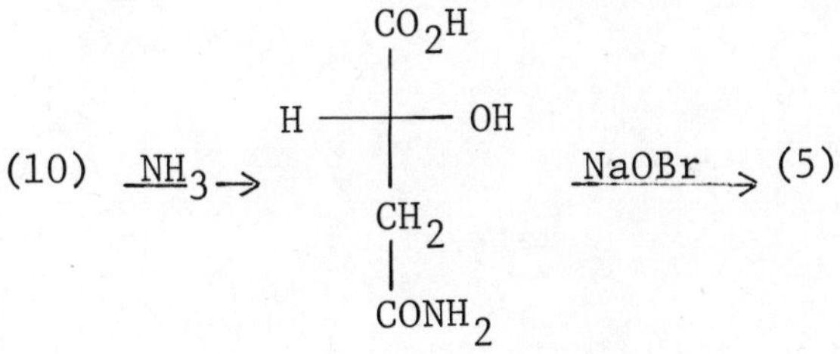

$$(10) \xrightarrow{NH_3} \begin{array}{c} CO_2H \\ H - C - OH \\ CH_2 \\ CONH_2 \end{array} \xrightarrow{NaOBr} (5)$$

Q-41 What is the configuration of (+)-tartaric acid (6) relative to that of (+)-glyceraldehyde (1)? (Hint: Use the correlations already worked out.)

CO_2H H — OH HO — H CO_2H	CHO H — OH CH_2OH
(+)-tartaric acid (6)	(+)-glyceraldehyde (1)

A-41

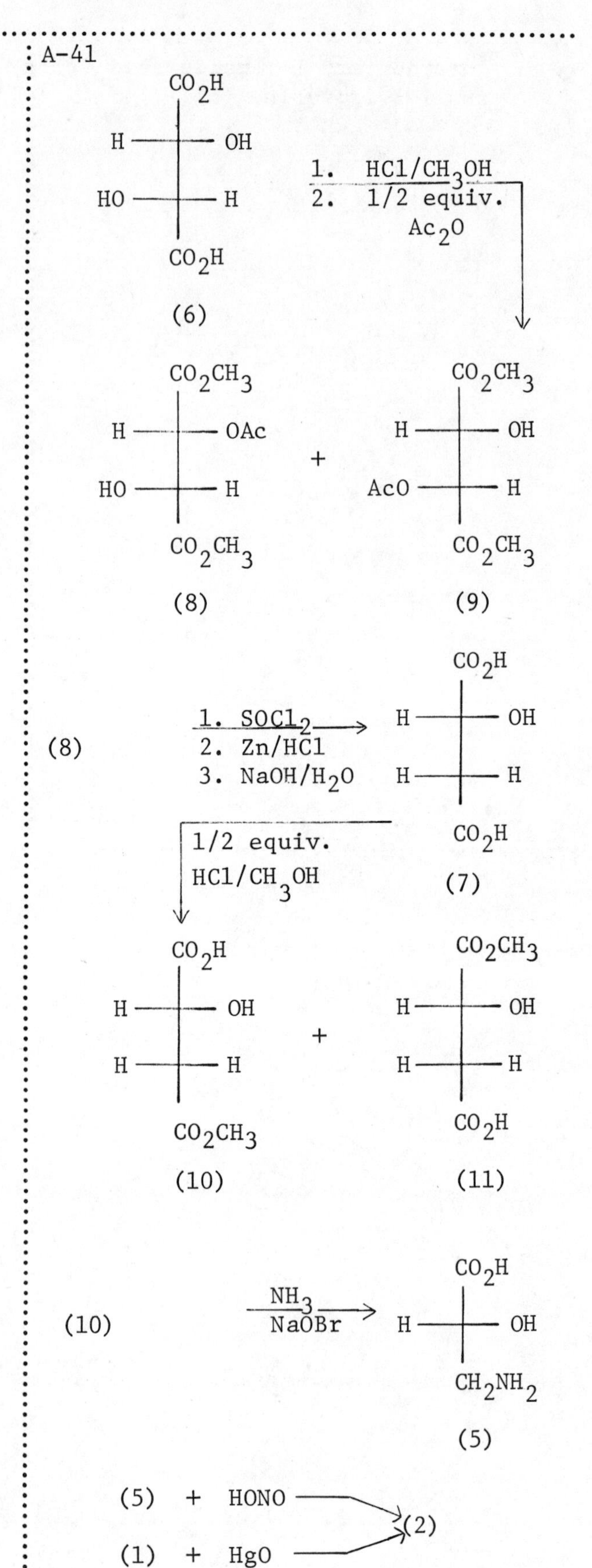

Q-42 Correlate the configuration of (+)-tartaric acid (6) with (-)-lactic acid. Use correlations already worked out, plus one additional reaction.

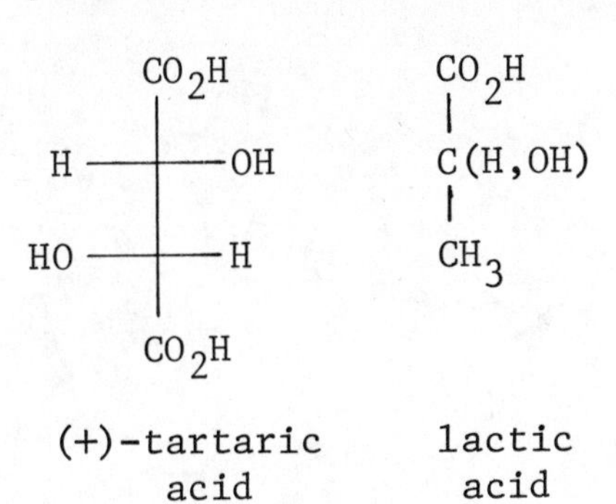

(+)-tartaric acid

(6)

lactic acid

A-42 As given in A-39 and A-40

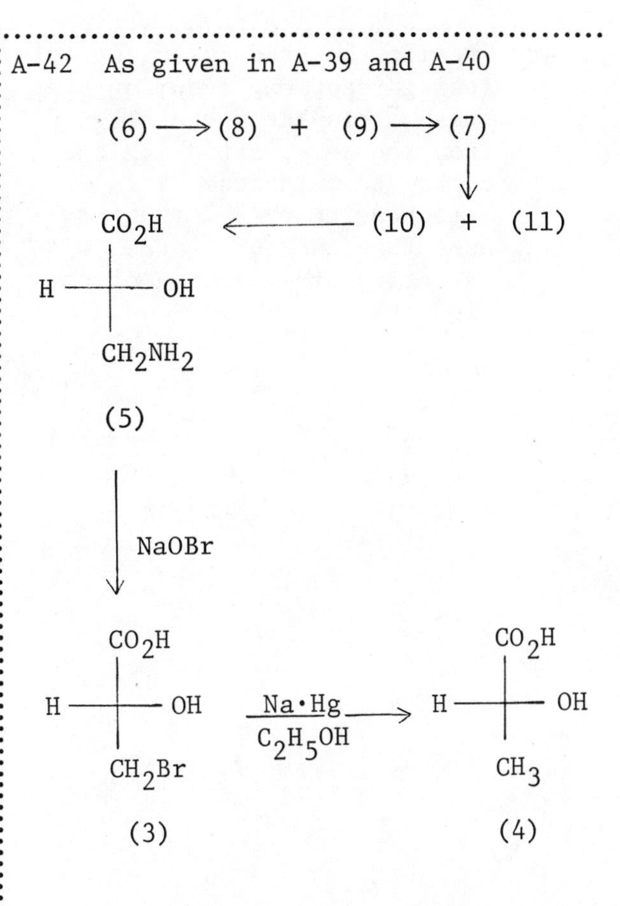

If (4) is levorotatory (-), then (+)-tartaric acid (6) and (-)-lactic acid have the same relative configuration. (In fact, (4) is the (-)-enantiomer of lactic acid.)

Relative Configurations II: S_N1, S_N2, and Intramolecular Rearrangements.

S-5 Correlations of relative configuration of compounds with asymmetric carbon atoms sometimes can be carried out by chemical interconversions in which the bonds to the asymmetric carbon atoms are broken. The three ways of doing this are S_N2 reactions, S_N1 reactions, and intramolecular rearrangements at the asymmetric carbon atoms.

Q-43 Predict the product of the following reaction, assuming that the N_3^- displaces the Cl group from the back, attacking the carbon to which the Cl is attached, in an S_N2 process. Does the reaction proceed with retention or inversion of configuration?

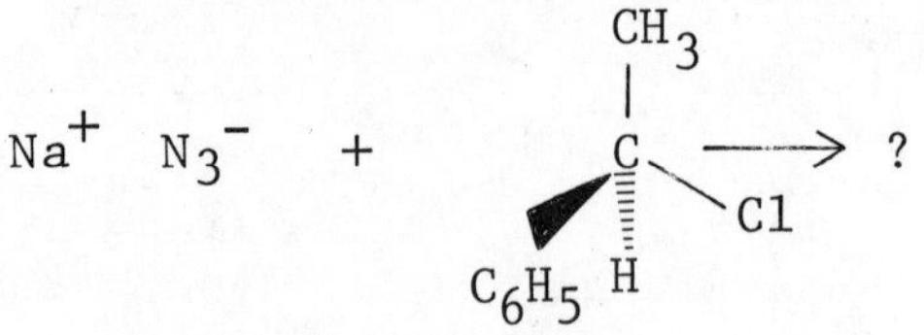

$Na^+\ N_3^-$ + (C with CH_3, Cl, C_6H_5, H) ⟶ ?

A-43

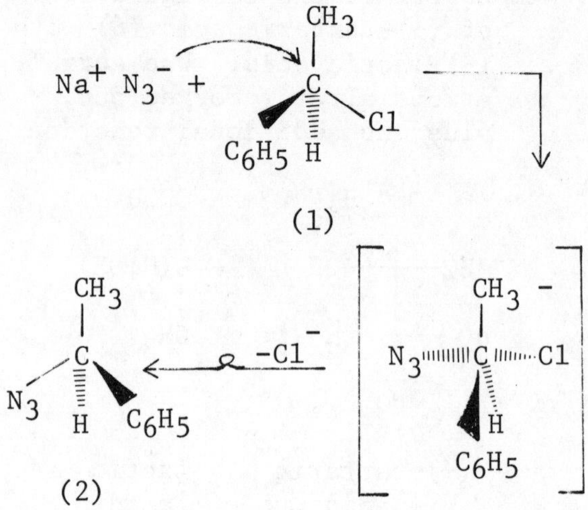

$Na^+\ N_3^-$ + (C with CH_3, Cl, C_6H_5, H)

(1)

[N_3 ··· C ··· Cl, with CH_3, H, C_6H_5]$^-$ —$-Cl^-$→ (C with CH_3, N_3, H, C_6H_5)

(2)

In an S_N2 mechanism, a nucleophilic substitution which obeys bimolecular kinetics occurs at a saturated carbon atom. The groups on the saturated carbon are inverted like the ribs of an umbrella in a strong wind. The inversion of configuration which occurs in an S_N2 mechanism is indicated with a curled arrow.

S_N2 Rule: Substitution by an S_N2 mechanism results in inversion of configuration.

Q-44 Will the product of this reaction be racemic?

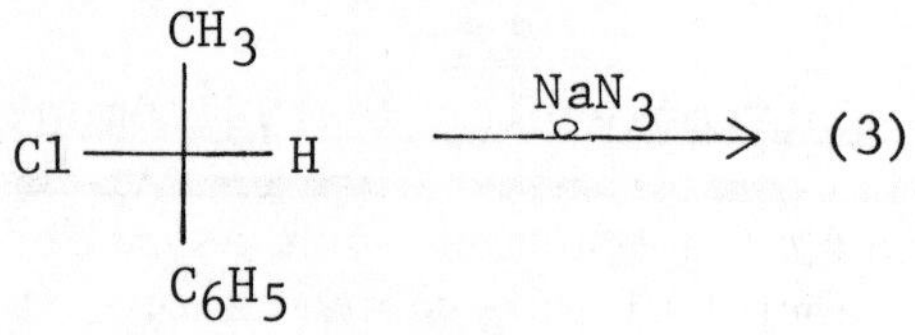

Cl—C(—CH_3)(—C_6H_5)—H $\xrightarrow{NaN_3}$ (3)

(+)-α-phenethyl chloride (1)

A-44 No, the product is optically active and has a configuration inverted relative to the starting material.

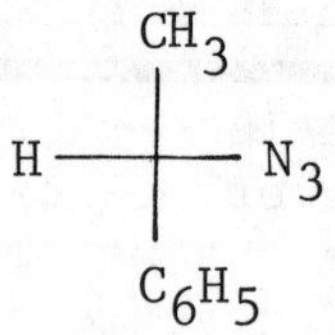

H—C(—CH_3)(—C_6H_5)—N_3

(3)

Q-45 Optically active 2-octyliodide can be racemized by treatment with iodide ion. Suggest a mechanism.

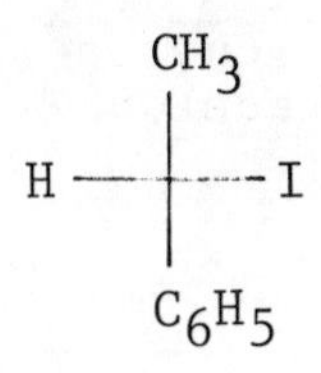

A-45 A displacement of iodide by iodide occurs and proceeds with inversion. The result is an equilibrium mixture in which equal amounts of each enantiomer are present.

$$\text{H}-\underset{\text{C}_6\text{H}_{13}}{\overset{\text{CH}_3}{\text{C}}}-\text{I} \;\underset{\text{I}^-}{\overset{\text{I}^-}{\rightleftharpoons}}\; \text{I}-\underset{\text{C}_6\text{H}_{13}}{\overset{\text{CH}_3}{\text{C}}}-\text{H}$$

Q-46 What would be required to prove that the following reaction goes with inversion of configuration as shown?

$$\text{Br}-\underset{\text{CH}_3}{\overset{\text{CO}_2\text{H}}{\text{C}}}-\text{H} \xrightarrow{\text{NH}_3} \text{H}-\underset{\text{CH}_3}{\overset{\text{CO}_2\text{H}}{\text{C}}}-\text{NH}_2$$

A-46 Any reaction which occurs by an S_N2 mechanism proceeds with inversion of configuration. To prove an S_N2 mechanism, one must demonstrate that a substitution which obeys bimolecular kinetics has occurred at a saturated carbon atom.

Since the given reaction already is shown to be a substitution at a saturated carbon, it is only necessary to show in addition that the reaction obeys bimolecular kinetics (second order kinetics).

Q-47 Predict the product of the following S_N1 reaction, assuming an intermediate carbonium ion forms which is attacked by the water. Does the reaction proceed with retention or inversion of configuration?

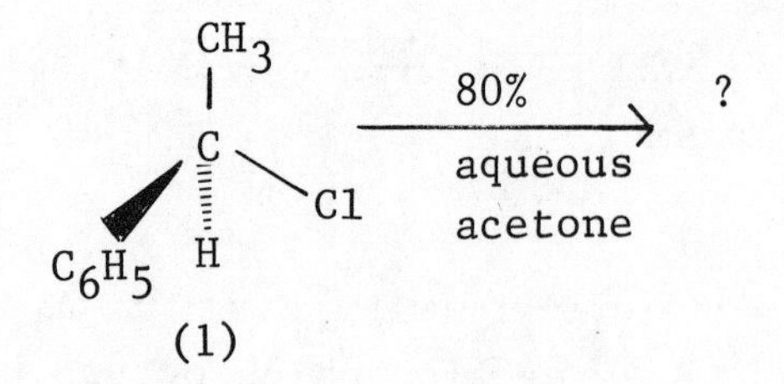

A-47

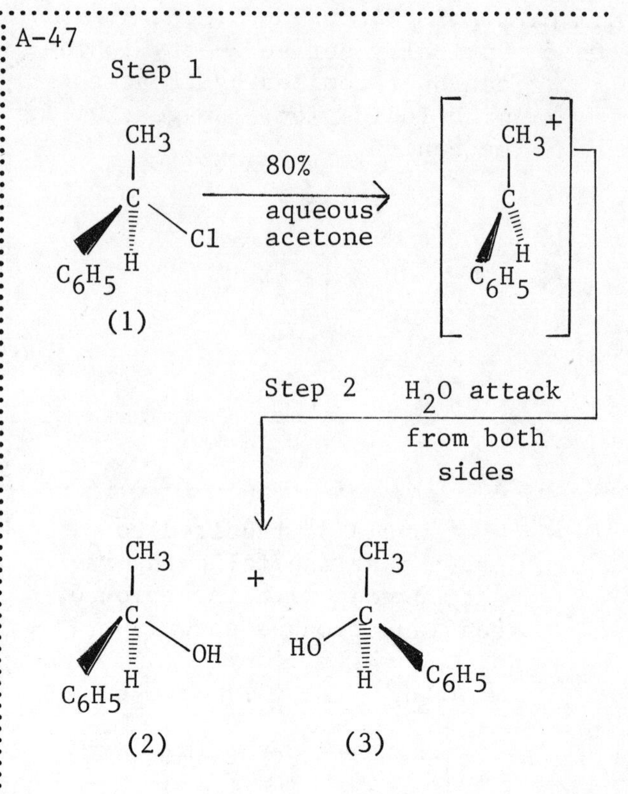

98% racemic product
2% net inversion

The reaction proceeds with both retention and inversion of configuration, that is, it is mostly a racemization reaction.

In the S_N1 mechanism (substitution, nucleophilic, unimolecular) substituion occurs in two steps. In the first step, the compound ionizes, forming a carbonium ion. In the second step, this carbonium ion reacts with a nucleophile. The kinetics are usually first order with the rate depending only on the concentration of substrate, because the first step, the ionization, is usually the rate-determining step. The racemization that frequently occurs results from the presence of a flat, symmetrical carbonium ion intermediate in the reaction.

S_N1 Rule: Substitution by an S_N1 mechanism usually gives complete racemization or racemization with some excess of inversion.

Q-48 In the S_N1 reaction,

(1) → (2) + (3). What relative percentages of (2) and (3) are obtained? Hint: % (2) + % (3) = 100%.

A-48 % (3) + % (2) = 100%
% (3) - % (2) = 2%
Therefore,
% (2) = 49%
% (3) = 51%

98% racemic product = all of (2) + 49% of (3)

Q-49 Draw the product of this S_N1 reaction. Is the product optically active? Explain.

CH_3 (top), Cl (left), H (right), C_6H_5 (bottom) $\xrightarrow{AlCl_3}$

A-49 The product is not optically active; it is racemic because the reaction mechanism is S_N1.

(4) $\underset{Cl^-}{\overset{AlCl_3}{\rightleftharpoons}}$ [CH_3, C, H, C_6H_5] + $AlCl_4^-$ ← $AlCl_3$

Cl^-

(5) CH_3 (top), H (left), Cl (right), C_6H_5 (bottom)

Q-50 What would one expect from the following reaction under conditions which do not provide complete second order kinetics (some first order kinetics are mixed in)?

CH_3 (top), Cl (left), H (right), C_6H_5 (bottom) $\xrightarrow{NaN_3}$?

(4)

A-50 One would expect racemization with some excess of inversion. This would afford both products (7) and (6), but more (7) than (6).

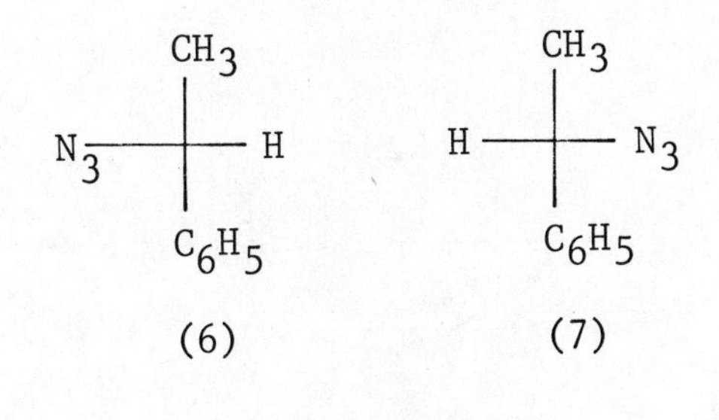

Q-51 Although kinetic studies have shown that this reaction proceeds by an S_N1 mechanism, retention of configuration occurs.

$$NH_3^+ - C(CO_2^-)(CH_3) - H \longrightarrow Br - C(CO_2^-)(CH_3) - H$$

(8)

How might the α -carboxylate group have preserved the configuration?

Hint: $RN^+H_3 \xrightarrow{NOBr} RN_2^+Br \dashrightarrow R^+ + N_2 + Br^-$

A-51 The α -carboxylate group forms an " α -lactone" type of intermediate which only allows the Br to come from that side which results in retention of configuration.

$$(8) \longrightarrow [\text{α-lactone-type intermediate: } {}^-O, C{=}O, C^+, CH_3, H] \longleftarrow Br^- \dashrightarrow (9)$$

Q-52 Offer an explanation for the observation that NH_3 gives retention with (10) but inversion with (11).

$$Br - C(CO_2H)(CH(CH_3)_2) - H \xrightarrow{NH_3} NH_2 - C(CO_2H)(CH(CH_3)_2) - H$$

(10)

$$Br - C(CONH_2)(CH(CH_3)_2) - H \xrightarrow{NH_3} H - C(CONH_2)(CH(CH_3)_2) - NH_2$$

(11)

A-52 The α -carboxylate in the first reaction helps preserve the configuration as discussed in A-51, while the α -amide group in the second reaction is unable to react in the same manner. The second reaction is S_N2.

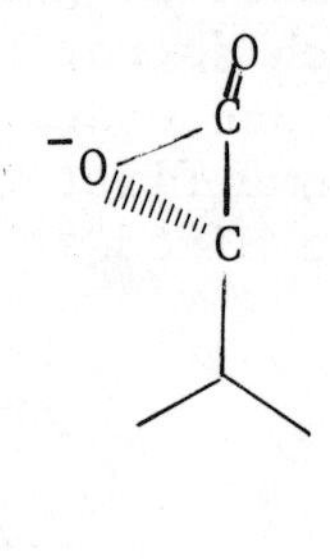

Q-53 In the Hofmann degradation shown below an unsubstituted amide is converted into an amine by an aqueous solution of bromine plus sodium hydroxide and then aqueous acid. Supply the missing intermediate (2) and show how it converts to the isocyanate intermediate (3).

A-53 The R group undergoes an intramolecular 1, 2 shift in the nitrene intermediate (2). This rearrangement may be represented as follows.

$$R{-}C({=}O){-}\ddot{N}: \longrightarrow (3)$$

$$NaOH + Br_2 \rightleftharpoons NaBr + HO{-}Br$$

$$RCONH_2 \quad (1)$$

$$RCO\overset{+}{N}HBr{-}H \xrightarrow{OH^-} RCO\ddot{N}(Br){-}H \xrightarrow{OH^-} (2)$$

$$(2) \longrightarrow O{=}C{=}\ddot{N}{-}R \ (3) \xrightarrow{H_3O^+} (HO)(H{-}O)C{=}N{-}R \xrightarrow{H^+} H{-}O{-}C({=}O){-}NHR \xrightarrow[-CO_2]{H^+} RNH_2 \ (4)$$

Q-54 What is required mechanistically in order for an optically pure amine (4) to result from an optically pure starting amide (1) when the R group is attached by an asymmetric carbon? (Two possibilities.)

A-54 The R group must migrate with either complete retention or inversion of configuration.

Q-55 Which do you think is most likely? Explain.

A-55 You are entitled to your own opinions and analysis. The fact is that the migration occurs with retention of configuration. Rearrangements involving intramolecular 1, 2 shifts of asymmetric carbons occur with retention of configuration.

Q-56 Draw the product of the reaction shown below. Is the product optically active? Explain.

$$\begin{array}{c} CONH_2 \\ | \\ CH_3 - \!\!\!+\!\!\! - H \\ | \\ C_6H_5 \end{array} \xrightarrow{NaOH, Br_2} \ ?$$

(-)

A-56

$$\begin{array}{c} NH_2 \\ | \\ CH_3 - \!\!\!+\!\!\! - H \\ | \\ C_6H_5 \end{array}$$

(-)- α -Phenethylamine

Yes, the product of this Hofmann degradation is optically active because the starting material is optically active. The rearrangement proceeds with retention of configuration.

Q-57 The Baeyer-Villiger reaction involves oxidative rearrangement of a ketone to an ester by a per-acid such as perbenzoic acid. Show how the C_2H_5 group in intermediate (2) can migrate to afford the ester (3). What is the leaving group?

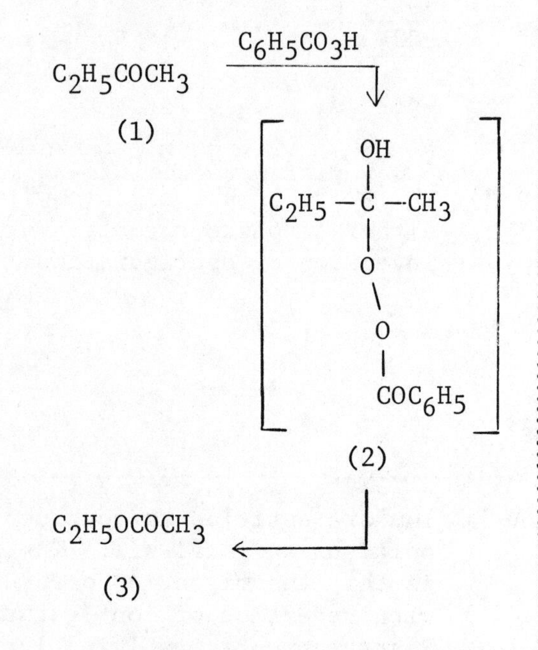

A-57

$$\begin{array}{c} H \\ | \\ O \\ | \\ C_2H_5 - C - CH_3 \\ | \\ O \\ \backslash \\ O \\ | \\ COC_6H_5 \end{array} \longrightarrow (3)$$

The leaving group is <u>a</u> benzoate ion, $C_6H_5CO_2^-$.

Q-58 Give the structures of compounds (5) and (6), assuming retention of configuration for any intramolecular 1, 2 shifts.

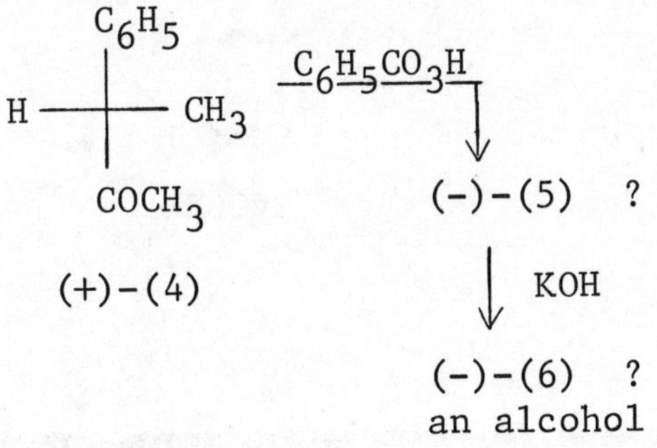

$$\underset{(+)\text{-}(4)}{\text{H}-\overset{C_6H_5}{\underset{COCH_3}{C}}-\text{CH}_3} \xrightarrow{C_6H_5CO_3H} (-)\text{-}(5)\ ? \xrightarrow{KOH} (-)\text{-}(6)\ ?$$

(−)-(6) ? an alcohol

A-58 In the first reaction, a Baeyer-Villiger oxidative rearrangement gives (5). In the second reaction, the KOH hydrolyzes (5) to yield (−)-phenylmethylcarbinol (6).

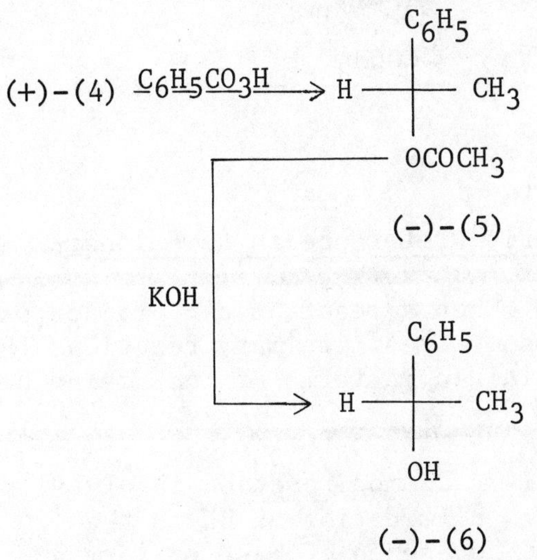

$$(+)\text{-}(4) \xrightarrow{C_6H_5CO_3H} \underset{(-)\text{-}(5)}{\text{H}-\overset{C_6H_5}{\underset{OCOCH_3}{C}}-\text{CH}_3} \xrightarrow{KOH} \underset{(-)\text{-}(6)}{\text{H}-\overset{C_6H_5}{\underset{OH}{C}}-\text{CH}_3}$$

Q-59 In both Q-57 and Q-58, a methyl group could also have migrated. But the principal observed products are as shown. In fact, the migratory aptitude in the alkyl series is tertiary > secondary > primary. In the aryl series, it is $p\text{-MeOC}_6H_5$ > C_6H_4 > $p\text{-NO}_2C_6H_4$. Suggest a reason for the observation that the groups with a greater capacity for electron release have the greater migratory aptitude

A-59 Evidently the oxygen to which the migratory group transfers becomes electron deficient prior to the migration, perhaps as indicated below, so that this oxygen is captured most rapidly by whichever adjacent group, R_1 or R_2, is more electron releasing.

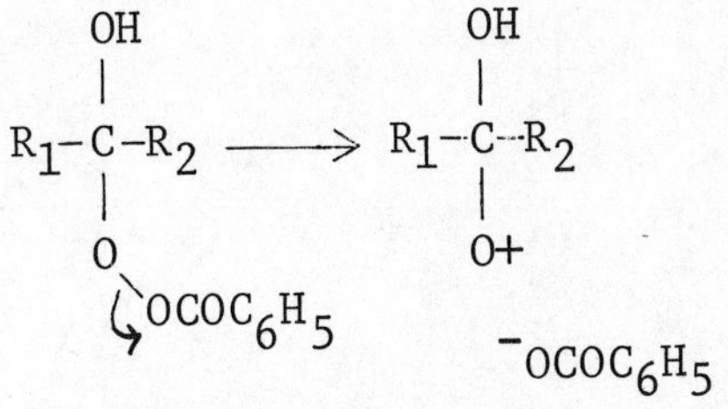

$$R_1-\overset{OH}{\underset{O-OCOC_6H_5}{C}}-R_2 \longrightarrow R_1-\overset{OH}{\underset{O^+}{C}}-R_2 \quad {}^-OCOC_6H_5$$

Q-60 Predict the main product of this reaction

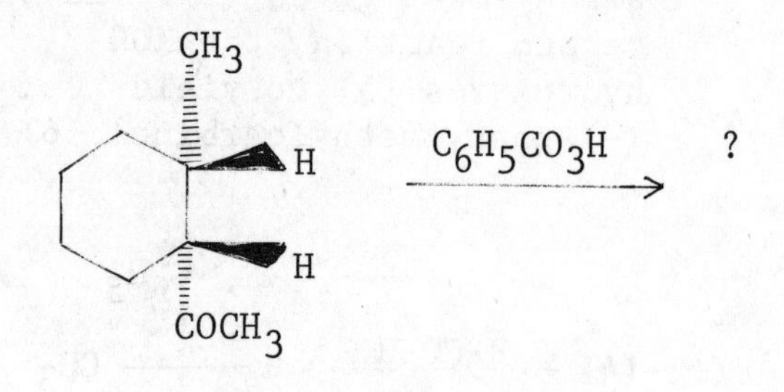

A-60 The tertiary group has a higher migratory aptitude than the primary methyl group.

CH3
H
H
OCOCH3

Steric Preference in Reactions:

S-6 When a reaction can produce stereoisomeric products, often one product predominates. In many reactions the preference for one product can best be explained in terms of the stereochemistry of one or more reactants.

Q-61 A common reaction involves elimination of HX in the presence of base to form an alkene. This reaction follows a characteristic stereochemical pattern. How can you explain the following results?

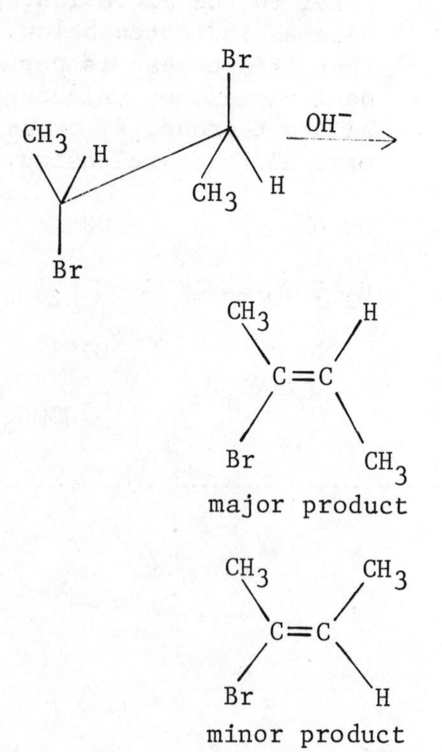

major product

minor product

A-61 The reactive conformation is one in which a hydrogen atom on one carbon is *trans-* to a bromine atom on the other carbon:

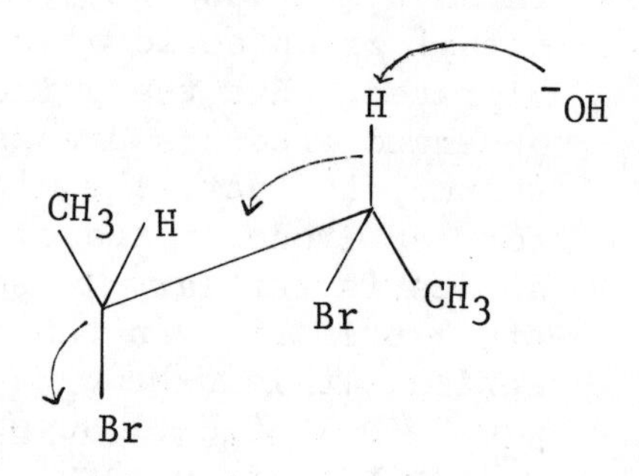

This type of reaction mechanism is called a *trans-* elimination.

Note: the same major product results from *trans-* elimination of the other HBr.

Q-62 What is the principal product of this reaction?

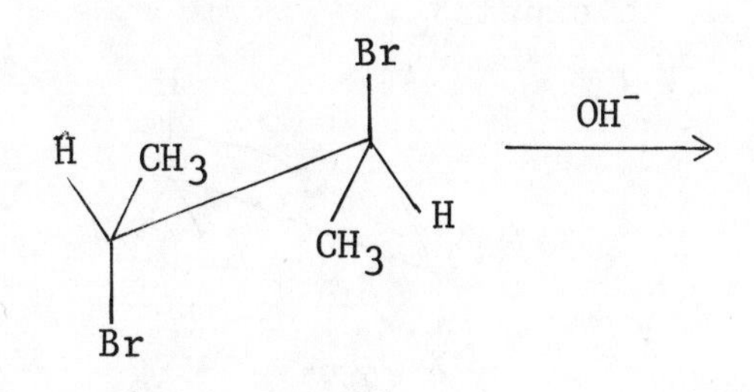

A-62 Here again the product is formed in a *trans*- elimination (again, *trans*- loss of other HBr gives same product).

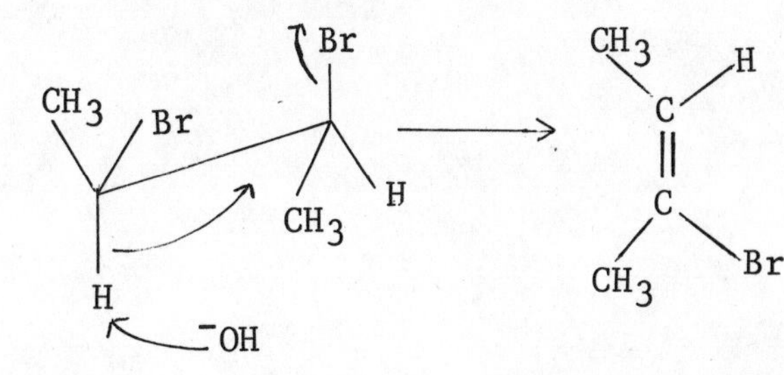

Q-63 Reaction of A with base to form an alkene is much slower than the reaction of B. Explain.

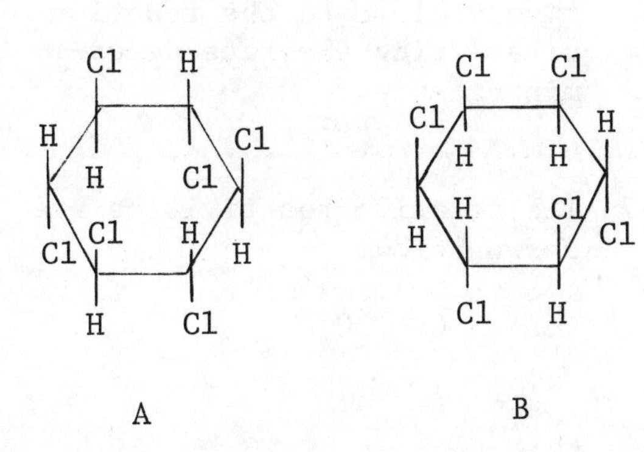

A-63 In A there is no possibility for *trans*- elimination of HCl. However there are four different ways that B can undergo *trans*-elimination (shown by arrows).

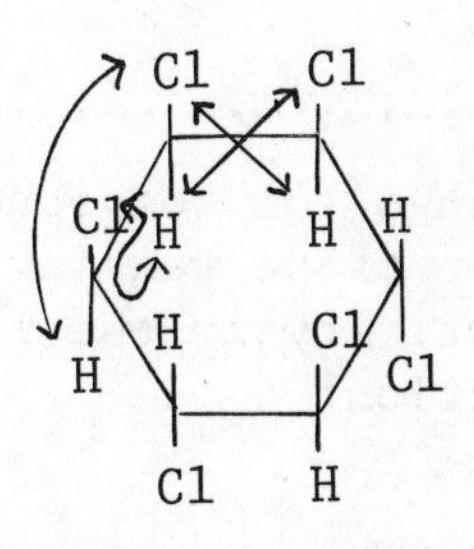

Q-64 Which of the two conformers of *cis*-2-chloromethylcyclohexane will be more reactive to dehydrohalogenation with base?

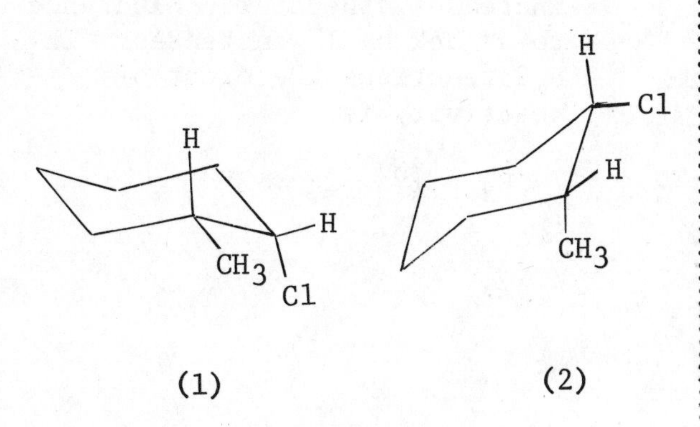

A-64 Conformer 1, in which the hydrogen and chlorine are in axial positions, is much more reactive. In cyclic systems the leaving groups must be in a diaxial conformation, as well as *trans*- to each other, for facile elimination to occur.

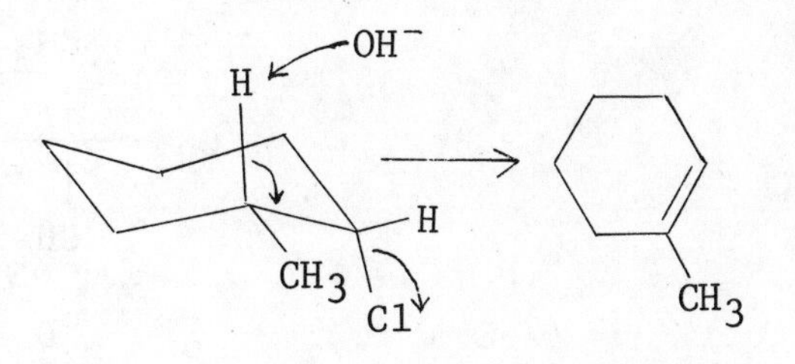

Q-65 The *trans*- elimination is is mechanistically similar to an S_N2 reaction. Explain.

A-65 In *trans*- elimination the electron pair in the C-H bond is released and attacks the adjacent carbon from the back side to expel X^-.

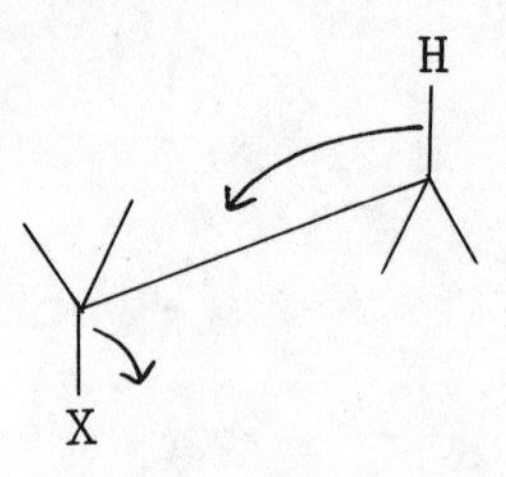

Q-66 In many situations the size of substituents near the site of reaction affect the course of a reaction. Would you expect S_N1 or S_N2 to be more seriously affected by this factor?

A-66 S_N2 reactions would be more affected by the size of substituents near the site of reaction. In an S_N2 reaction the nucleophile must approach relatively close to the reaction site during the rate-determining step.

Q-67 Arrange the compounds below in order of decreasing S_N2 reactivity in the following reaction:

$$RBr + I^- \longrightarrow RI + X^-$$

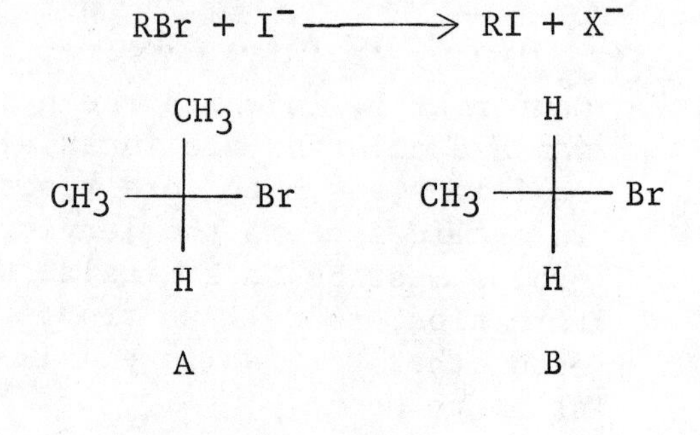

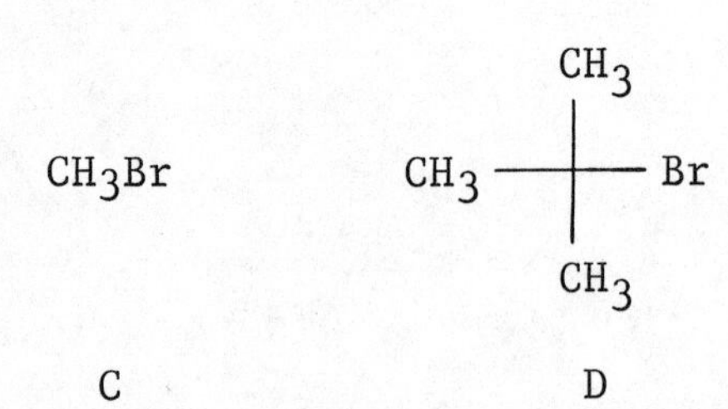

Explain your reasoning.

A-67 The relative reactivities are shown below:

C	30
B	1
A	0.01
D	0.001

As the size of the substituents, increases, the steric hindrance to attack by I^- increases. In S_N2 reactions the order of reactivity is

$$CH_3 > 1^o > 2^o > 3^o$$

Q-68 The following relative reactivities seem to violate the general order of relative reactivities given in the previous answer. Explain the apparent contradiction.

		relative rate
$(CH_3)_2CHX$	(2°)	0.02
$(CH_3)_3CX$	(3°)	≈0
$(CH_3)_3C\text{-}CH_2X$	(1°)	0.00001
neopentyl halide		

A-68 The neopentyl halide reacts much more slowly than expected for a primary halide because of the very great steric hindrance caused by the adjacent *t*-butyl group.

Q-69 Which product would you expect to predominate in the following reaction? Why?

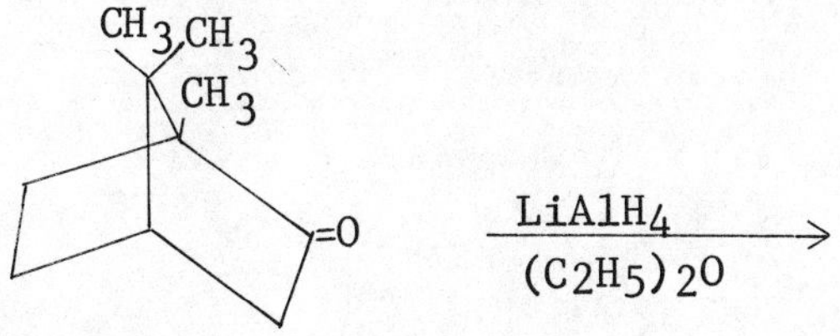

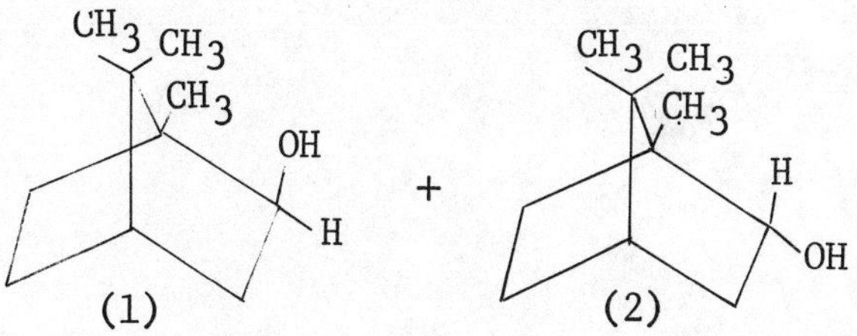

CH_3 CH_3 CH_3 =O $\xrightarrow[(C_2H_5)_2O]{LiAlH_4}$

CH_3 CH_3 CH_3 OH H (1) + CH_3 CH_3 CH_3 H OH (2)

A-69 Compound 1 which results from attack from the less hindered underside should predominate. The relative yields actually obtained are:

Compound	Yield
1	90%
2	10%

Q-70 Explain the following product distribution:

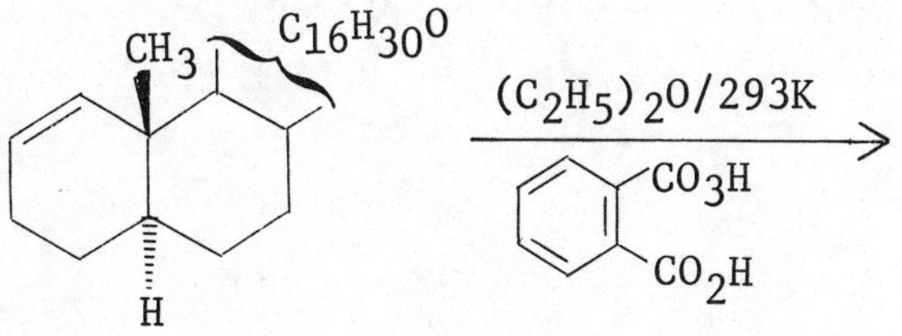

CH_3 $C_{16}H_{30}O$ H $\xrightarrow{(C_2H_5)_2O/293K}$ (phthalic monoperacid: CO_3H, CO_2H)

H O CH_3 $C_{16}H_{30}O$ H major product

O H CH_3 $C_{16}H_{30}O$ H minor product

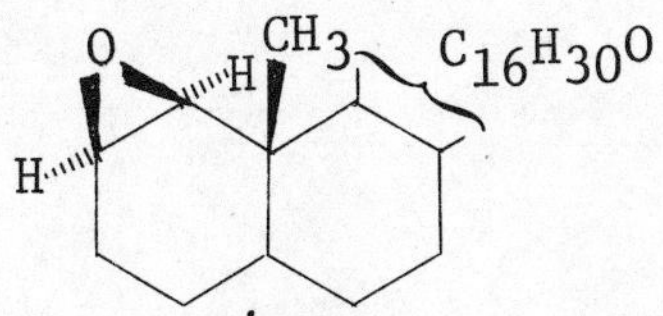

A-70 Evidently the peracid attacks the olefin from the less hindered side.

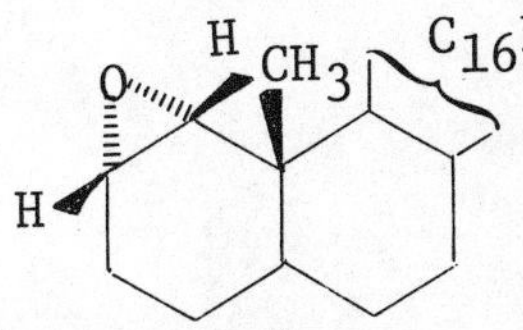

Q-71 Explain the following product distribution:

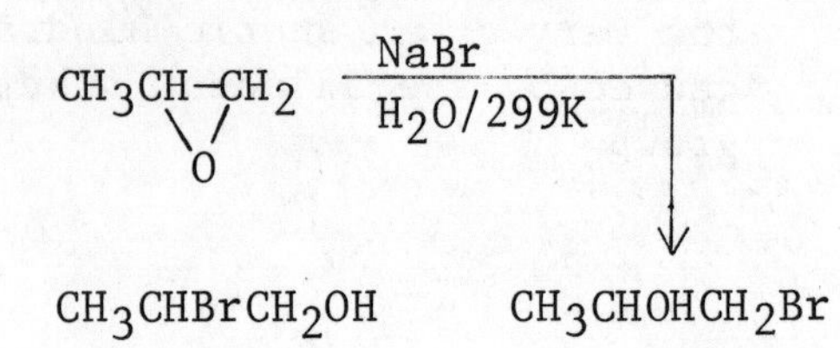

$CH_3CHBrCH_2OH$ 5% $CH_3CHOHCH_2Br$ 95%

What type of mechanism is involved?

A-71 Since the major product has resulted from attack at the less hindered carbon, the mechanism is probably S_N2.

Q-72 Boron hydrides react with olefins to form compounds which can be converted to alcohols by treatment with hydrogen peroxide. In the example below which product would you expect to predominate?

$CH{=}CH_2$ (on cyclohexene ring)

1. $(CH_3CHCH_3CHCH_3)_2BH$
2. H_2O_2

(1) CH_2CH_2OH (on cyclohexene ring) (2) $CH{=}CH_2$, HO (on cyclohexane ring) (3) $CH{=}CH_2$, OH (on cyclohexane ring)

A-72 Since the boron hydride reagent is rather bulky, reaction at the less hindered double bond is favored. The major product is (1).

Q-73 Explain the product distribution in the following reaction. Consider the compound to be locked into the conformation shown by the necessity for the bulky t-butyl group to be equatorial.

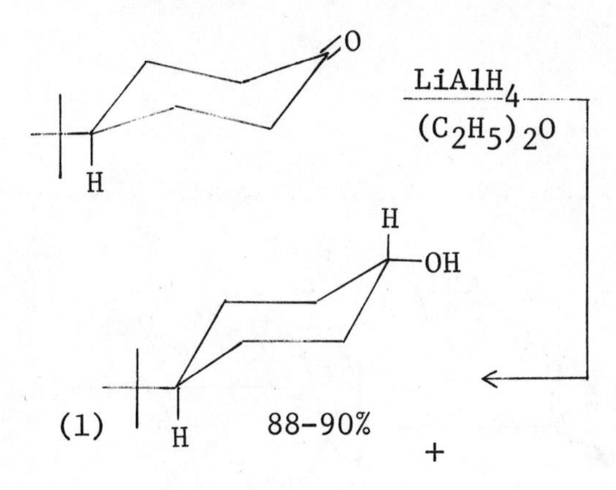

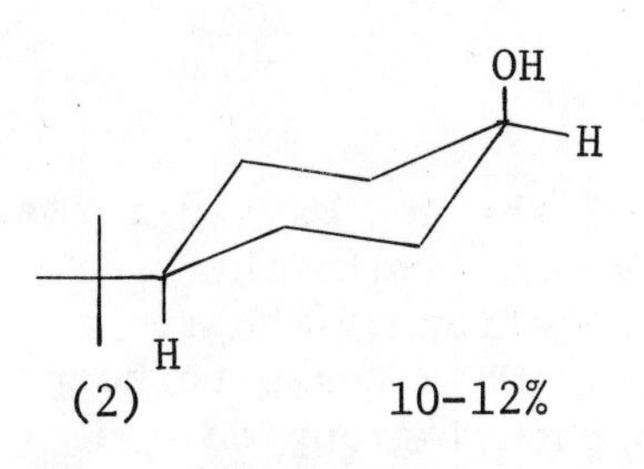

(Hint: Which product would be favored if the product distribution were controlled by the steric properties of the reactants?)

A-73 The major product is the less hindered one. In this case the steric properties of the product are more important. The transition states leading to the two products are shown below:

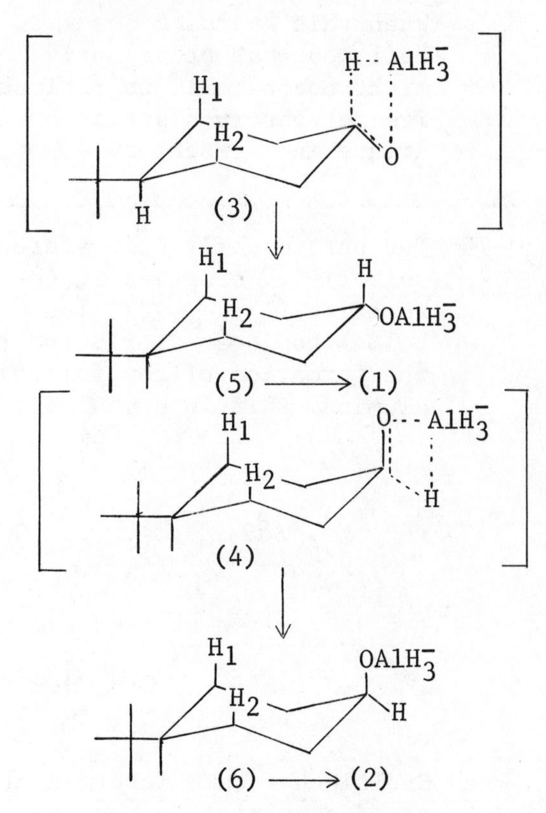

Steric interference between H_1 and H_2 in (3) and the approaching metal hydride (steric approach control) will oppose the formation of the equatorial alcohol (1). On the other hand, steric interference between the developing alkoxyaluminum group and H_1 and H_2 in (4) (product development control) will oppose the formation of the axial alcohol (2). Since (2) is the minor product, apparently the latter steric interference is of greater magnitude, that is, the reaction undergoes product development control.

Q-74 Sometimes the steric preference of a reaction is best explained based on the stereochemistry of one or more of the *reactants*.

When this approach seems to fail, on what other basis might one be able to rationalize any observed steric preference in a reaction?

A-74 On the basis of the stereochemistry of one or more of the *intermediates* involved in the reaction.

Q-75 The particularly high stereoselectivity observed in the reaction (1) ⟶ (2) (see below) has been attributed to the formation of the following aluminum salt intermediate.

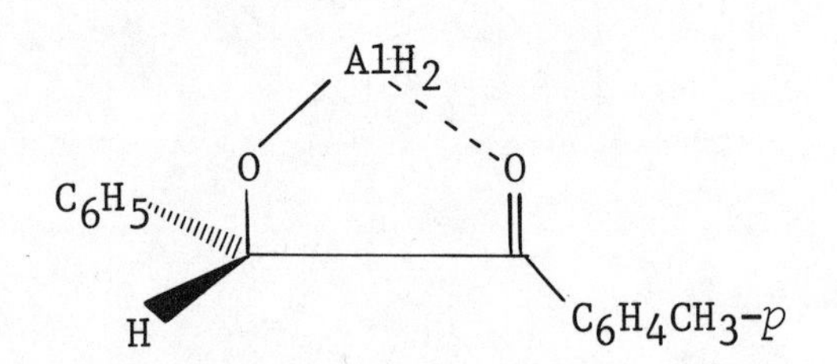

Explain how this intermediate could give rise to the observed steric preference in the reaction.

$C_6H_5CHOHCO(p\text{-}CH_3C_6H_4)$ $\xrightarrow{LiAlH_4}$

(1)

OH OH H

C_6H_5 H p-$CH_3C_6H_4$

(2)

A-75

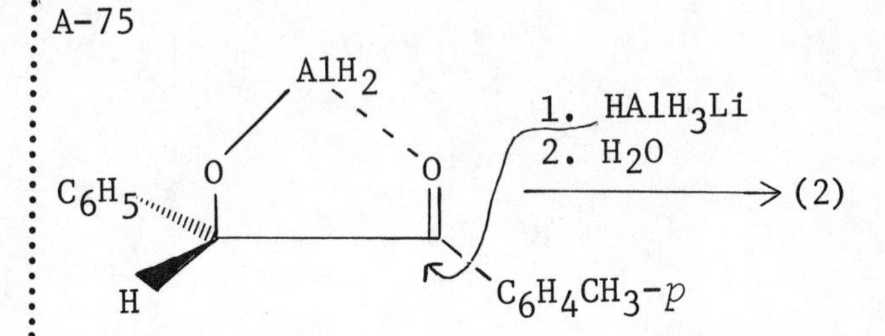

By bridging the two oxygen atoms, the AlH_2 group limits the internal rotation of the reactant (1). This tends to keep the bulky phenyl group on one side only of the carbonyl which limits the accessibility of the carbonyl on that side. Consequently, the carbonyl is attacked by hydride primarily from the other less-hindered side as shown. Stereoisomer (2), the major product, results from this mode of attack.

Q-76 Alkaline peroxidation of *trans*-3-methyl-pent-3-ene-2-one, (1), gives only the epoxide (2).

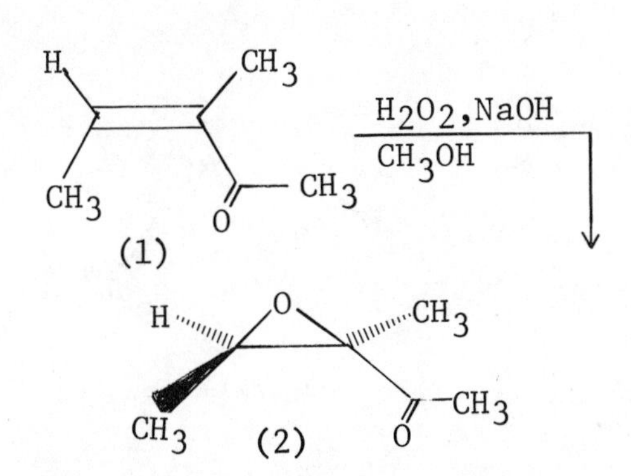

Does this reaction show steric preference? Why?

A-76 Yes, because (2) is a stereoisomeric product; the other possible stereoisomer of (2) is (3),

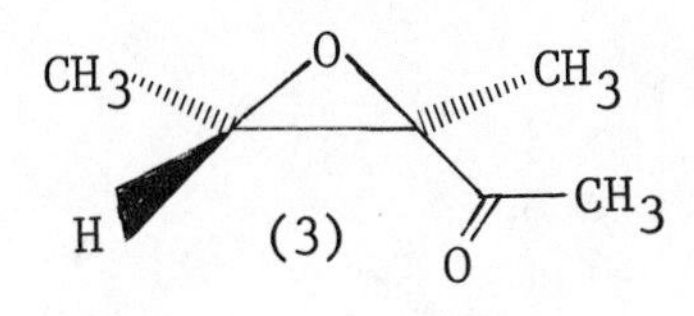

yet only (2) is observed.

Q-77 How might one determine whether the steric preference of (1) ⟶ (2) (of Q-76) is based on the stereochemistry of the reactant, (1), or on a subsequent intermediate?

A-77 One could carry out the reaction with the *cis*-enone (4) instead of the *trans*-enone (1).

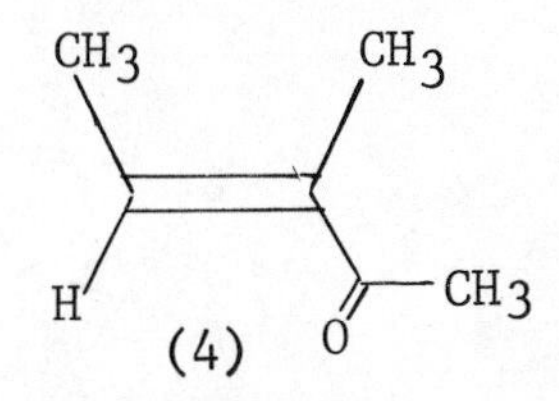

In fact, alkaline peroxidation of (4) also leads to (2).

Q-78 How is it possible for two different stereoisomers, (1) and (4), (of Q-76 and A-77) to give the same stereoisomer, (2)?

A-78 Both (1) and (4) must react to give a common *intermediate*, (5) which reacts further to become (2).

Q-79 How could the intermediate (5) (see A-78) react further to become (2) (see Q-76)?

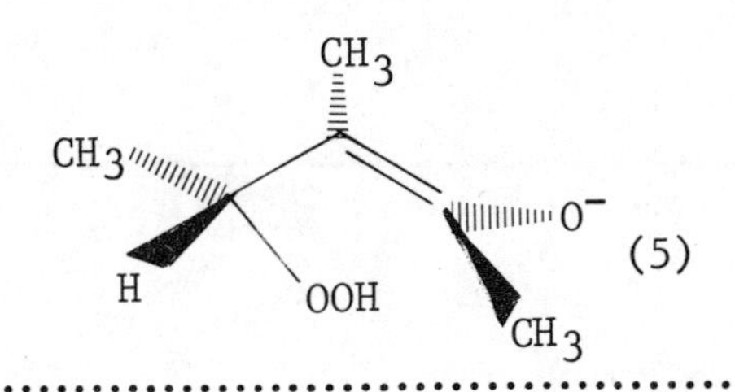

A-79 [(5)] ⟶ (2)

Q-80 Similar reaction of the intermediate (6) would afford the other stereoisomer, (7), which is not observed.

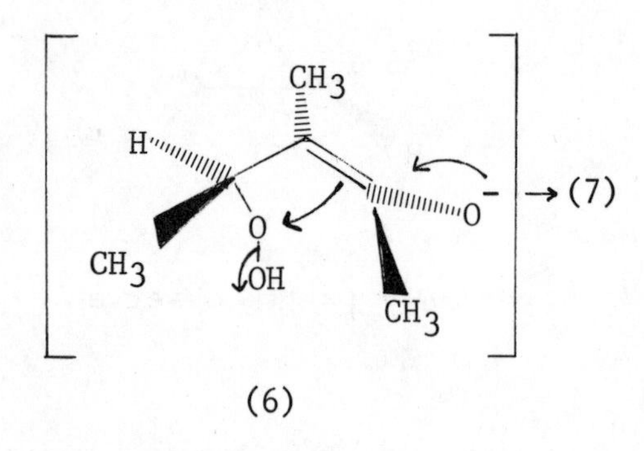

(6)

Explain why this intermediate is less favorable than intermediate (5).

A-80 Intermediate (6) is energetically less favorable because of steric hindrance between the two starred methyl groups.

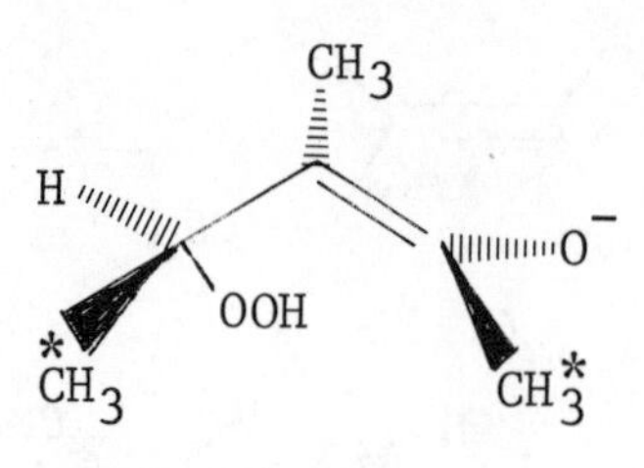

Q-81 Bromination of *trans*-2-butene, (1), results in *meso*-2,3-dibromobutane, (2).

(1) (2)

Reaction of the other stereoisomer of (1) with bromine results in the other stereoisomer of (2). Draw this reaction and name these other two compounds.

A-81 *Cis*-2-butene, (3), reacts with bromine to yield *dl*-2,3-dibromobutane (4A plus mirror image).

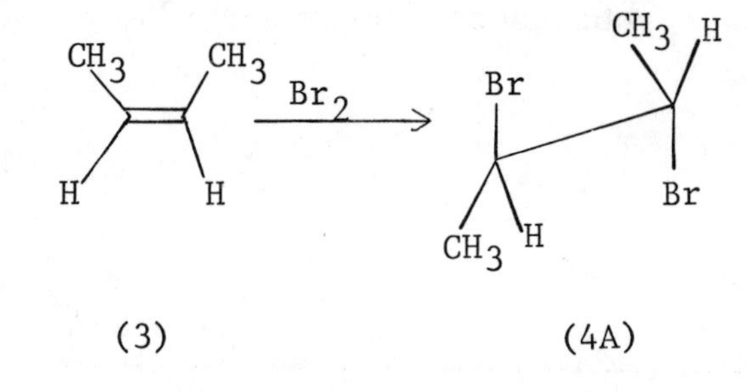

(3) (4A)

("*dl*" indicates that the product is racemic. See S-1, p. 99.)

Q-82 Does this suggest that the reactions (1) ⟶ (2) and (3) ⟶ (4) (see Q-81 and A-81) share a common intermediate stereoisomer?

A-82 Absolutely not, because the products (2) and (4) are different stereoisomers.

Q-83 Propose mechanisms to account for the steric preference observed in the bromine addition reactions given in Q-82.

A-83 Formation of stereoisomeric intermediate pi-complexes (bromonium ion intermediates) (5) and (6) which maintain the initial stereochemistry of their precursors, (1) and (3), while they undergo further *trans* attack of Br^- will account for the observed steric preference.

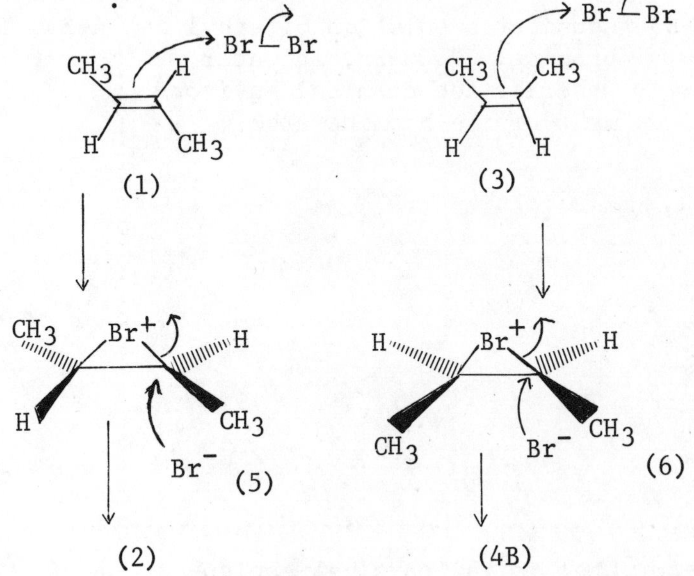

CH_3 H Br—Br H CH_3 (1)

CH_3 CH_3 Br—Br H H (3)

CH_3 Br^+ H H CH_3 Br^- (5)

H Br^+ H CH_3 CH_3 Br^- (6)

(2)

(4B)

Q-84 Offer an explanation for this steric preference.

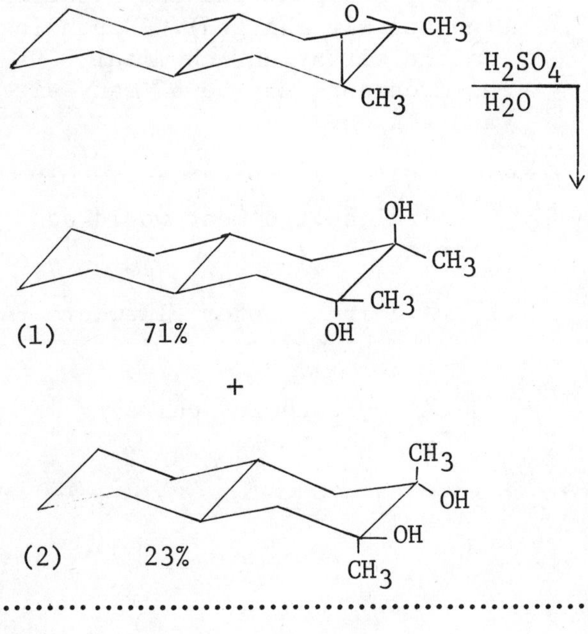

O CH_3 CH_3 $\xrightarrow[H_2O]{H_2SO_4}$

OH CH_3 CH_3 OH (1) 71%

+

CH_3 OH OH CH_3 (2) 23%

A-84 *Trans*-addition of water occurs. Attack of water occurs primarily as in (3) since this leads to a diequatorial orientation of the two bulky groups. The diequatorial orientation is more stable than a diaxial one.

(3) H O^+ CH_3 CH_3 OH_2 ⟶ (1)

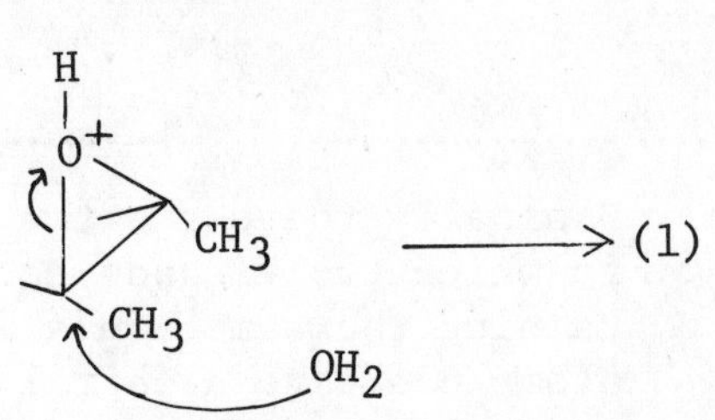

(4) H O^+ CH_3 CH_3 OH_2 ⟶ (2)

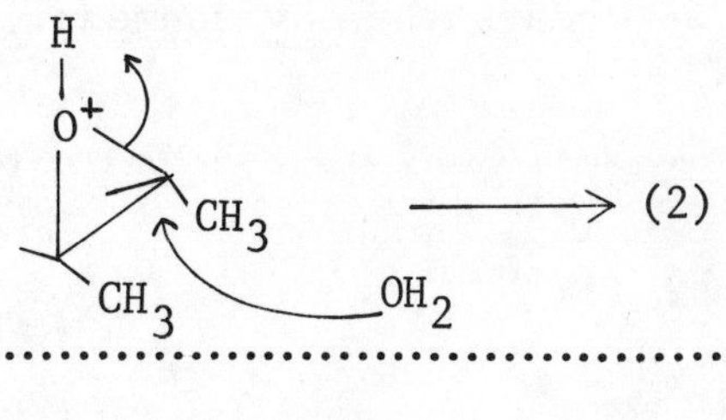

Chemical Environments:

S-7 In Section III it was noted that often two ligands in a molecule which are seemingly identical are in fact geometrically different. As a result the two ligands may have chemically different environments. If they do, they will have different chemical and physical properties. In general the chemical environment of a ligand is determined by the nature of its neighboring atoms and bonds. This includes the atoms and bonds of nearby solute and solvent molecules.

Q-85 Consider a solution of ethyl bromide, CH_3CH_2Br, in water. Describe the chemical environment of the bromine atom.

A-85 The chemical environment of the bromine atom results from the following factors:

a) Atoms and bonds in the same molecule: five hydrogen and two carbon atoms, seven bonds.

b) Solvent molecules containing hydrogen and oxygen atoms and unshared electrons; H-O bonds.

c) Other solute molecules.

Q-86 What is the physical explanation of "chemical environments?"

A-86 These "environments" are the sum of the electrical fields associated with each atom and bond. The fields are determined by the magnitude, direction, and type of motions of the nuclei (positively charged) and the electrons (negatively charged). Depending on the situation, chemical environments may be of many sizes and shapes.

Q-87 Basically there are two ways in which two ligands containing the same bonds and atoms can have a chiral environment. What are they? (Hint: Recall what factors create the environment.)

A-87 A chiral environment would be created by:

a) A chiral center elsewhere in the molecule.

b) A non-racemic, chiral solvent.

Q-88 Under what circumstances will ligands which are equivalent experience different chemical environments?	A-88 Equivalent ligands will always experience the *same* chemical environment whether the environment is chiral or achiral.
Q-89 How could H_A and H_B in the following molecule be subjected to a chiral environment? 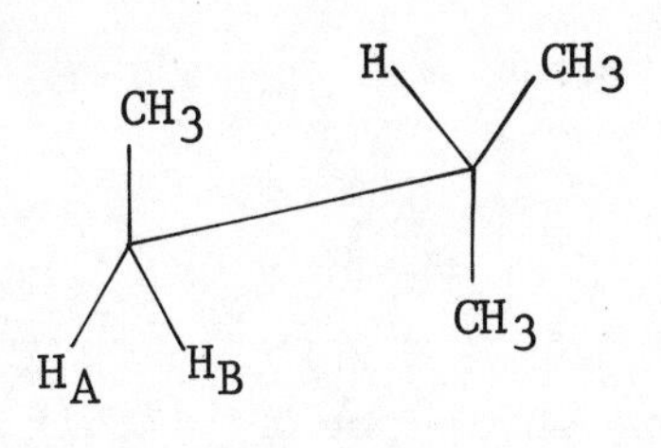	A-89 By dissolving the molecule in a non-racemic, chiral solvent. (The molecule itself has no chiral elements.)
Q-90 Would the chemical and physical properties of H_A and H_B be different when the above molecule is dissolved in a non-racemic, chiral solvent? Explain.	A-90 No, because H_A and H_B are equivalent.
Q-91 Enantiotopic ligands have the same chemical environment under achiral circumstances. Will their environments still be the same in a chiral situation?	A-91 No, enantiotopic ligands have have different environments under chiral circumstances.
Q-92 Diastereotopic ligands always have different chemical environments under both chiral and achiral circumstances. Why is it that enantiotopic ligands can be considered, in a sense, to become diastereotopic when placed in a non-racemic chiral medium?	A-92 The reaction of enantiomers with a chiral reagent leads to diastereomers which have different chemical and physical properties. In a similar way the interaction of a chiral medium with enantiotopic ligands can be considered to produce diastereotopic ligands which also have different chemical and physical properties.

Q-93 Under what circumstances will H_A and H_B in the following compound have different chemical and physical properties?

CH_3 H CH_3

Cl H_B H_A

(1)

A-93 Under all conditions, since H_A and H_B are diastereotopic.

Q-94 Are H_A and H_B in (1) situated in a chiral or achiral environment?

A-94 The environment is chiral since the molecule has a chiral center (*).

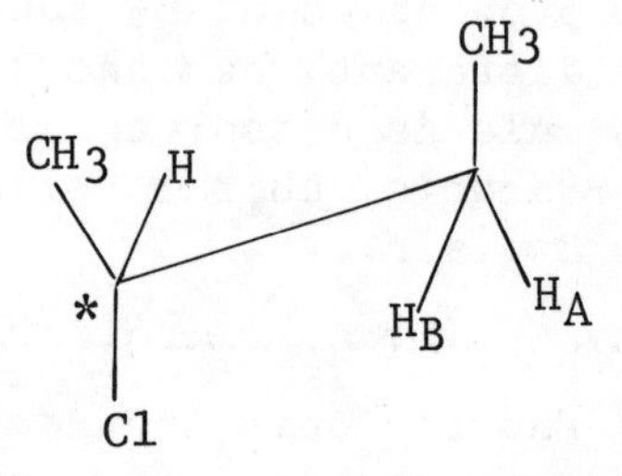

Q-95 Under what circumstances will the two CO_2H groups in this molecule have different chemical and physical properties? Explain.

H

CH_3 C CO_2H

CO_2H

A-95 Only when they are subjected to a chiral environment, as when the molecule is dissolved in a non-racemic solvent. The two CO_2H groups are enantiotopic.

Differentiation of Like Ligands in Chemical Reactions:

S-8 Ligands which have the same atoms and bonds but exist in different chemical environments, as noted previously, can be expected to have different chemical properties. Therefore, the two like ligands should behave differently in chemical reactions. In other words, they are differentiable chemically. This frequently occurs in biological reactions.

Q-96 The four compounds shown below are common reactants in biological reactions. Each compound has two like CH_2 groups. In which cases are the two CH_2 groups in the compounds differentiable?

CH_2CO_2H
HO — C — CO_2H
CH_2CO_2H

citric acid

CO_2H
H — C — H
H — C — H
CO_2H

succinic acid

CH_2OH
C=O
CH_2OH

dihydroxyacetone

CH_2OH
H — C — OH
CH_2OH

glycerol

A-96

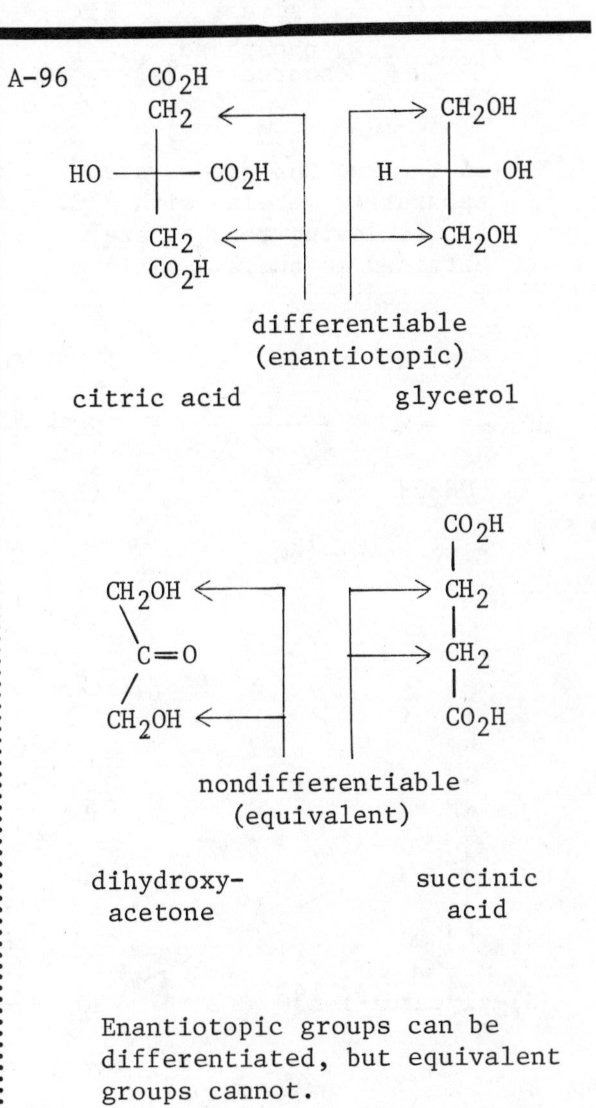

Enantiotopic groups can be differentiated, but equivalent groups cannot.

Q-97 In the presence of an appropriate phosphate source, glycerol is converted to glycerol phosphate by the enzyme glycerol kinase:

$$\begin{array}{c} CH_2OH \\ H-\!\!\!\!\!-\!\!\!\!\!+\!\!\!\!\!-\!\!\!\!\!-OH \\ CH_2OH \end{array} \xrightarrow[\text{phosphate source (ATP)}]{\text{glycerol kinase}} \begin{array}{c} CH_2OPO_3H_2 \\ H-\!\!\!\!\!-\!\!\!\!\!+\!\!\!\!\!-\!\!\!\!\!-OH \\ CH_2OH \end{array}$$

If the two CH_2 groups are separately labeled with ^{14}C, the following results are obtained in the enzymatic reaction:

(R)-glycerol-1-^{14}C (Fischer projection: top CH_2OH, left HO, right H, bottom $^*CH_2OH$) → not converted to (1); converted to (2).

(1): top CH_2OP, left HO, right H, bottom $^*CH_2OH$

(2): top CH_2OH, left HO, right H, bottom $^*CH_2OP$

(S)-glycerol-1-^{14}C (Fischer projection: top $^*CH_2OH$, left HO, right H, bottom CH_2OH) → converted to (3); not converted to (4).

(3): top $^*CH_2OH$, left HO, right H, bottom CH_2OP

(4): top $^*CH_2OP$, left HO, right H, bottom CH_2OH

Note: P = PO_3H_2

What are the stereochemical relationships among (1), (2), (3), and (4)? (Assume no chemical difference between ^{12}C and ^{14}C.)

A-97 Compounds (1) and (4) represent one enantiomer of glycerol phosphate, compounds (2) and (3) the other.

Note that ^{14}C is preferred over ^{12}C in the R, S convention.

Q-98 Compounds (2) and (3), while differing in the position of the ^{14}C label, are both L- α -glycerol phosphate. Explain why only one enantiomer was produced in the reaction.

A-98 The two CH_2OH groups of glycerol are enantiotopic. Since the enzyme glycerol kinase is a chiral reagent, the two groups will have different chemical environments and thus be differentiable chemically.

Q-99 In the case of glycerol one CH_2OH group reacts much more rapidly than the other in the example given. How can you specify which group is the more reactive without making a drawing?

A-99 The *pro*-R CH_2OH group is the more reactive one.

Q-100 What are the like ligands in erythritol?

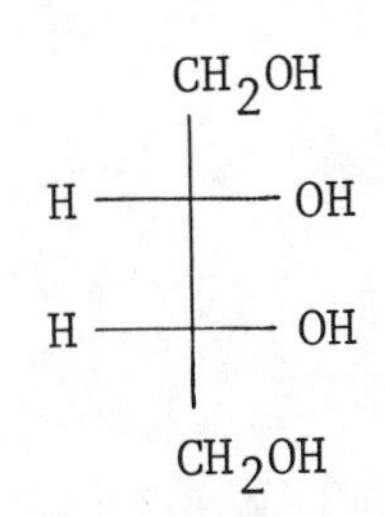

A-100 The like ligands are:

a) The terminal CH_2OH groups

b) The secondary OH groups

c) The methylene hydrogens

d) The tertiary hydrogens

Q-101 What is the nature of the stereochemical relationships existing between like ligands in erythritol?

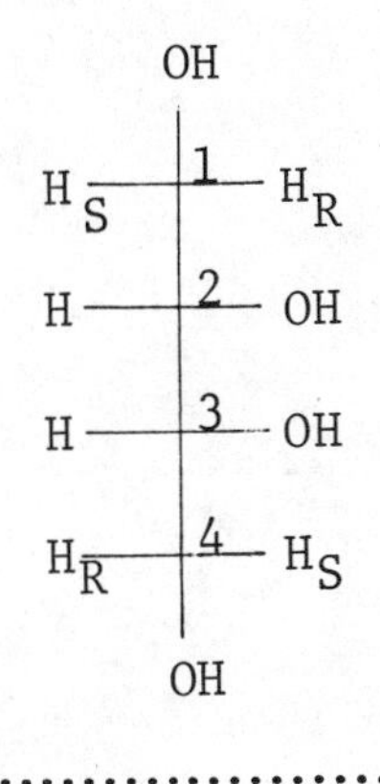

A-101 The stereochemical relationships are as follows:

a) The secondary OH groups- enantiotopic
b) The tertiary hydrogens- enantiotopic
c) The primary OH groups- enantiotopic
d) H_S at C-1: enantiotopic to H_R at C-4; diastereotopic to H_R at C-1 and H_S at C-4
e) H_R at C-1: enantiotopic to H_S at C-4; diastereotopic to H_S at C-1 and H_R at C-4

Q-102 Is a chiral agent needed to differentiate between the paired CH_2 hydrogens of a given CH_2OH group in erythritol?

A-102 No, since the two hydrogens are diastereotopic only a selective achiral chemical reagent is required.

Q-103 Can the terminal CH_2OH groups of L-threitol be differentiated chemically?

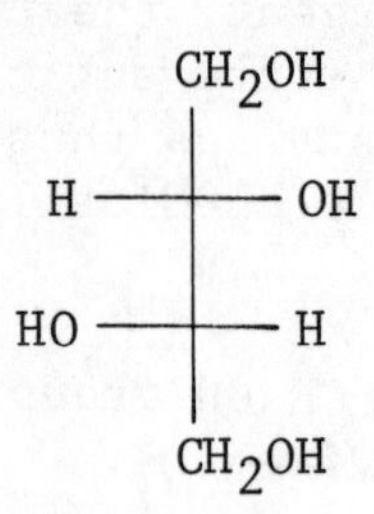

L-threitol

A-103 No, they are equivalent. In fact this has been proven experimentally.

Q-104 Explain why the two products in the following reaction are not produced in equal amounts.

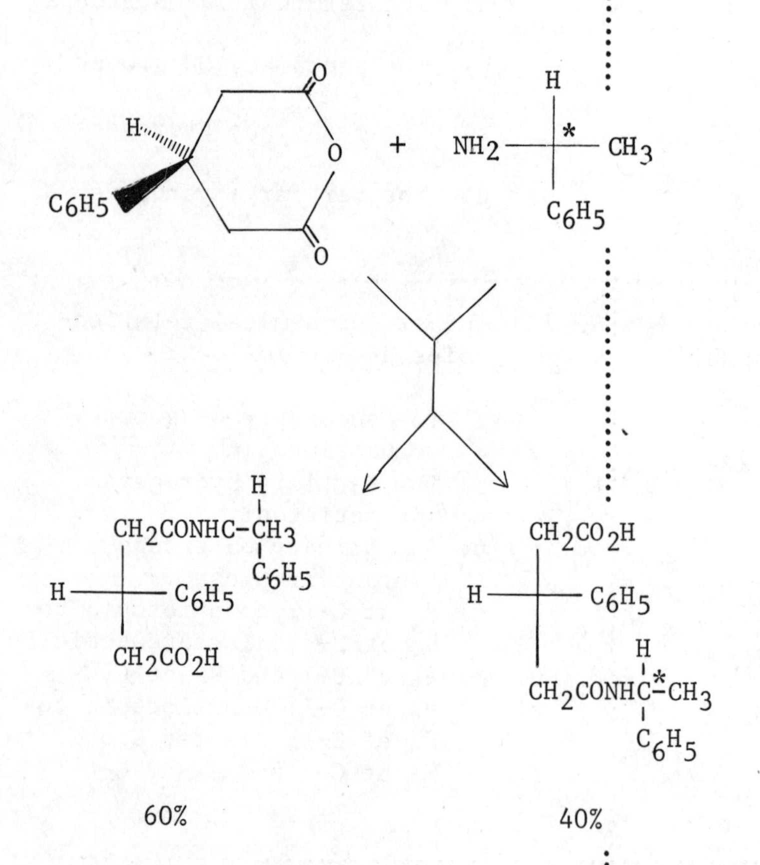

* = chiral center

A-104 The two CO groups are enantiotopic. Interaction with the chiral amine produces a chiral environment, allowing the CO groups to be differentiated.

Appendices

Equivalent, Enantiotopic, Diastereotopic

Type of Group	Substitution Criterion*	Symmetry Criterion	Physico-chemical Properties
Equivalent	No isomers generated	C_n $(n > 1)$ interchanges the groups	Always same
Enantiotopic	Enantiomers generated	S_n $(n > 1)$ interchanges the groups	Same only in achiral medium
Diastereotopic	Diastereomers generated	No symmetry operation exchanges groups	Always different

* Substitute with an achiral group, e.g., a halogen atom.

Symmetry Symbols

The symmetry elements and operations are represented by symbols. Each symmetry element shares a common symbol with its corresponding symmetry operation.

Symmetry Element	Symmetry Operation	Symbol
Center (Point)	Reflect each atom through this point	i
n-Fold Cylindrical or Rotation Axis (Line)	Rotate the molecule by 360/n degrees about this axis*	C_n
n-Fold Rotation-Reflection Axis (Line)	Rotate the molecule by 360/n degrees about this axis and then reflect each atom through a plane perpendicular to this axis*	S_n
Plane (Plane)	Reflect each atom through this plane.	σ

*C_1 and S_1 (C_n and S_n when $n = 1$) are excluded because they correspond to rotation by 360°, which is not useful in stereochemistry.

To evaluate how symmetrical a molecule is, that is, what symmetry elements a molecule possesses, consider the molecule in its most symmetric conformation. Any molecule which possesses one or more symmetry elements (excluding C_1 and S_1) is defined to be symmetrical.

Reagents for Correlating Configurations

The following chemical reagents have been useful in chemical correlations of configuration. They have been provided to help you answer some of the questions in this section.

R = alkyl or aryl

To Convert This	To This	Use This
1. RCO_2H	RCO_2R'	$R'OH$, HCl
	RCH_2OH	$LiAlH_4$
2. RCO_2R'	RCO_2H	NaOH, then HCl
	$RCONH_2$	NH_3
	$RC(OH)R'_2$	$R'MgI$
3. RCH_2CONH_2	RCH_2NH_2	NaOBr
	RCH_2COR'	$R'MgI$
4. $RCOR'$	$RCHOHR'$	$NaBH_4$
5. RCHO	RCO_2H	HgO
6. RCH_2OH	RCH_2OCOCH_3	Ac_2O
	RCH_2Cl	$SOCl_2$
	RCH_2OR'	$R'I$, Ag_2O
	RCH_3	TsCl, then $LiAlH_4$
	RCH_2Br	PBr_3
7. RCH_2NH_2	RCH_2OH	HONO
	RCH_2Br	NOBr
8. RCH_2Br	RCH_3	Na·Hg, EtOH
9. $RCHClR'$	$RCHOHR'$	Zn/HCl

Mirror Image Classification of Stereoisomers

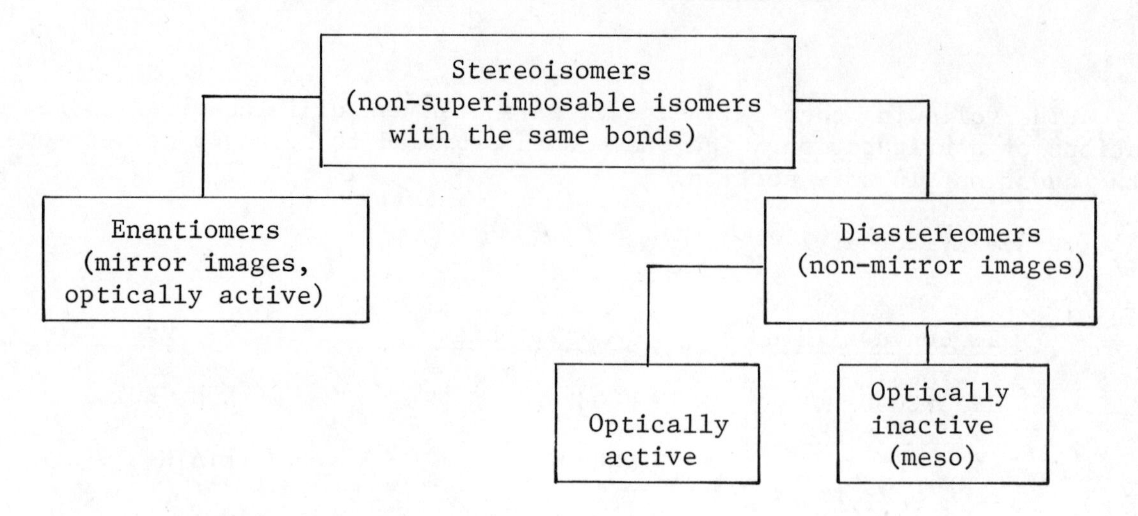

Symmetry Classification of Stereoisomers

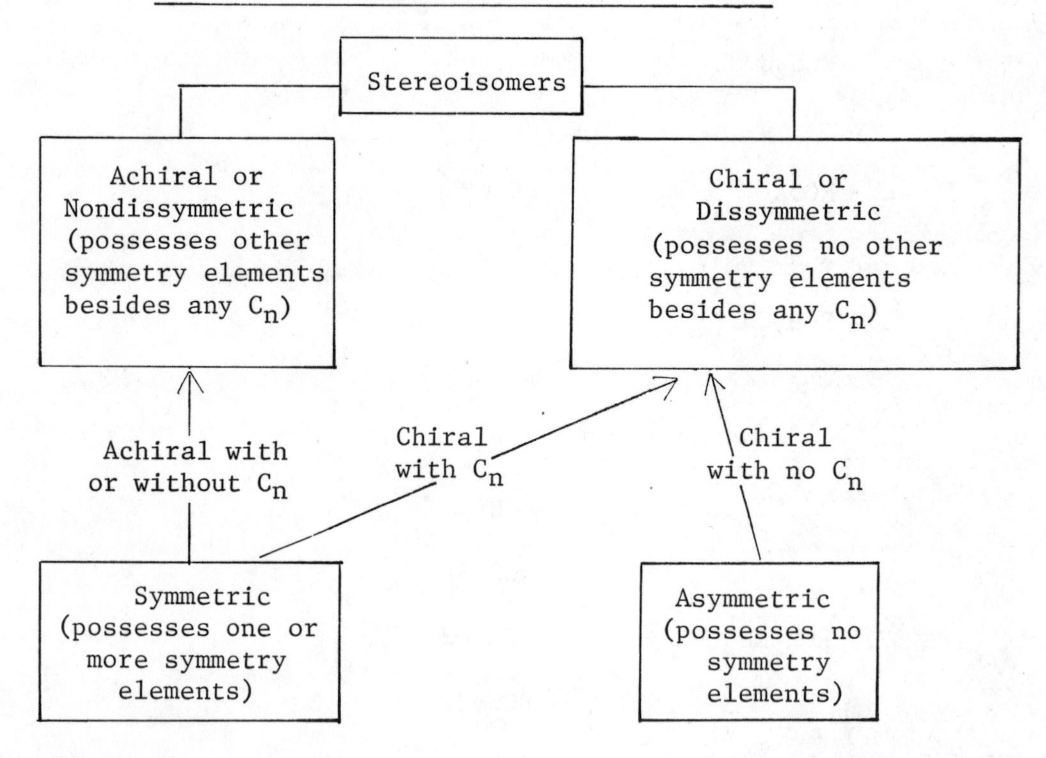

Note: Chirality, mirror images, optical activity, and symmetry are interrelated on pages 49-61.

Bibliography

This bibliography provides references for further reading in the main areas of stereochemistry. It is by no means exhaustive. In particular, no references have been included to articles which are highly technical or in journals which are not likely to be readily available.

GENERAL ARTICLES AND TEXTS

Abernathy, J. L., "The Concept of Dissymmetric Worlds," *J. Chem. Educ.*, *49*, 455 (1972).

The concept of dissymmetric worlds is presented as a way to organize the information relating to optical isomerism.

Allinger, N. L., and Eliel, E. L., ed., *Topics in Stereochemistry*, *8*, Wiley-Interscience, New York, 1974.

This is the eighth in a series of yearly volumes which started in 1967. The emphasis is on recent advances in stereochemistry; with each chapter covering a particular topic. Each new volume provides a cumulative index for the entire series. Some of the chapters from this and the previous volumes are described individually in this bibliography.

Aylett, B. J., and Harris, M. M., ed., *Progress in Stereochemistry*, *IV*, Butterworths, London, 1969.

This is the most recent volume in this series. Some of the chapters in this and the previous volumes are described individually in this bibliography. Aylett and Harris also edited volume III (1962), while W. Klyne and P. B. D. de la Mare edited volume II (1958) and W. Klyne edited volume I (1954).

Boschke, F. "Stereochemistry I" and "Stereochemistry II," (Vol. 47 and 48, respectively, of *Topics in Current Chemistry*), Springer-Verlag, New York, 1974.

Molecular propellers, the helicity of variously twisted chains of atoms, and cyclopropanes are covered in Volume I, while computer oriented representations, stereospecificity in biology, and [2.2] paracyclophanes are covered in Volume II.

Eliel, E. L., *Stereochemistry of Carbon Compounds*, McGraw-Hill Book Company, New York, 1962.

The classic introduction to stereochemistry. It covers both structural and dynamic aspects and is still very useful even though published more than a decade ago.

Eliel, E. L., "Teaching Organic Stereochemistry," *J. Chem. Educ.*, *41*, 73 (1964).

The author's views on the most effective way to incorporate stereochemistry in a general chemistry course.

Mislow, K., *Introduction to Stereochemistry*, W. A. Benjamin, Inc., New York, 1965.

A short introduction to the conceptual basis of stereochemistry. Emphasis is on the fundamentals of structural stereochemistry, rather than on dynamic aspects.

Newman, M. S., ed., *Steric Effects in Organic Chemistry*, John Wiley and Sons, Inc., New York, 1956.

One of the pioneering works in modern stereochemistry. It summarizes the substantial amount of stereochemical information which was already available in the mid-1950's.

Natta, G., and Farina, M., *Stereochemistry*, translated from the Italian edition (Milano, 1968) by A. Dempster, Harper and Row, New York, 1973.

This introductory level book briefly covers a wide variety of stereochemical topics, and other topics as well, such as chemical resonance and the duplication of DNA. The book largely is written at the level of a freshman college chemistry text.

BIOCHEMISTRY

Alworth, W. L., *Stereochemistry and Its Application in Biochemistry*, Wiley-Interscience, New York, 1972.

An introduction to the relation between substrate symmetry and biological specificity. The first three chapters are devoted to some history, an introduction to symmetry elements, and definitions and designations pertaining to configurations. The other three chapters include 26 specific examples of the application of stereochemical analysis to biochemical problems.

Bentley, Ronald, *Molecular Asymmetry in Biology*, Vols. I and II, Academic Press, New York, 1969.

Two volumes covering the fundamentals of stereochemistry from both a chemical and a biological perspective. Volume I is devoted to the chemical aspects, while Volume II illustrates the application of stereochemical analysis to several specific types of biochemical reactions.

Bentley, Ronald, "Configurational and Conformational Aspects of Carbohydrate Chemistry," *Annual Review of Biochemistry*, *41*, 953 (1972).

Nomenclature, physical methods, unusual new compounds, mutarotation isomers, and stereochemistry of enzymatic reactions are covered.

Ferrier, R. J., "Configurational Analysis in Carbohydrate Chemistry," in *Progress in Stereochemistry*, 4, 43 (1969).

Physical methods (nmr, polarimetry, ir, x-ray and neutron diffraction, chromatography and electrophoresis), chemical methods, and enzymatic methods are presented.

Goodwin, T. W., "Prochirality in Biochemistry," *Essays Biochem.*, *9*, 103 (1973).

A review with 69 references on the ability of enzymes to deal stereospecifically with prochiral centers of biochemical compounds.

Popjak, G., "Stereospecificity of Enzyme Reactions," in Boyer, P. D., ed., *The Enzymes*, Vol. II, 3rd edition, Academic Press, New York, 1970, p. 115.

A comprehensive review of the stereospecificity of enzymic reactions. The reactions discussed are NAD-dependent oxidoreductases, malate synthesis, reactions of the citric acid cycle, reactions at isolated double bonds, reactions of polyprenyl biosynthesis, and coenzyme B_{12}-dependent reactions.

Smith, W. G., and York, J. L., "Stereochemistry of the Citric Acid Cycle," *J. Chem. Educ.*, *47*, 588 (1970).

Attention is drawn to certain common textbook errors in presentation of the stereochemistry of the citric acid cycle.

Stoddart, J. F., *Stereochemistry of Carbohydrates*, Wiley-Interscience, New York, 1971.

This book achieves much organization that is not available elsewhere about carbohydrates in an interesting manner for the general reader.

CONFORMATIONAL ANALYSIS AND CONFIGURATION

Allinger, N. L., "Conformational Analysis in the Elementary Organic Course," *J. Chem. Educ.*, *41*, 70 (1964).

A very brief summary which emphasizes cyclohexane derivatives.

Barton, D. H. R., "The Principles of Conformational Analysis," *Science, 169,* 539 (1970).

The lecture delivered by the author when he received the Nobel Prize in 1969. This article provides a brief survey of the principles and historical development of conformational analysis.

Brewster, J. H., "Assignment of Stereochemical Configuration by Chemical Methods," Chapter XVII in Weissberger, A., ed., *Techniques of Chemistry,* Vol. IV, 2nd ed., ed. K. W. Bentley and G. W. Kirby, Wiley-Interscience, New York, Part III, 1972, p. 1.

A thorough review covering the configurations of olefins, relative configurations, absolute configuration, and selected examples.

Chiurdoglu, G., ed., *Conformational Analysis,* Academic Press, New York, 1971.

A collection of the papers presented at the International Symposium on Conformational Analysis, held in Brussels in September 1969. The emphasis is on the use of nmr methods.

Eliel, E. L., Allinger, N. L. Angyal, S. J., and Morrison, G. A., *Conformational Analysis,* Wiley-Interscience, New York, 1965.

Emphasis is given to acyclic and ring systems, natural products, physical methods, and carbohydrates. Excellent coverage of basic principles.

Hanack, M., *Conformation Theory,* translated from the German manuscript by H. C. Neumann, Academic Press, New York, 1965.

A short historical survey is followed by discussions of cyclic, monocyclic, bicyclic, polycyclic, and heterocyclic compounds.

Klyne, W., and Buckingham, J., *An Atlas of Stereochemistry: Absolute Configuration of Organic Molecules,* Oxford University Press, New York, 1973.

A reference work which tabulates absolute configurations of more than 3000 compounds.

Lambert, J. B., "The Shapes of Organic Molecules," *Scientific American, 222:1,* 58 (January 1968).

A very readable, well-illustrated introduction to the concept of conformation in chemistry.

Liberles, A., Greenberg, A., and Eilers. J. E., "Attractive Steric Effects," *J. Chem. Educ., 50,* 676 (1973).

Discussion of cases in which stereochemistry cannot be explained on the basis of repulsive interactions between nonbonded atoms or groups.

Lowe, J. P., "The Barrier to Internal Rotation in Ethane," *Science*, *179*, 527 (1973).

A comparison of different theoretical approaches to explaining sources of barriers to internal rotation about single bonds. The author concludes that a qualitative, intuitively useful explanation is possible.

Price, C. C., "Some Stereochemical Principles from Polymers," *J. Chem. Educ.*, *50*, 744 (1973).

Polymers are used to illustrate several important basic principles relating chemical structure to properties, including: molecular chirality, the influence of chemical structure on molecular geometry and particularly molecular flexibility, and entropy and its effect on the properties of these molecules.

Wilen, S. H., *Tables of Resolving Agents and Optical Resolutions*, Eliel, E. L., ed., University of Notre Dame Press, Indiana, 1972.

This compilation serves as a supplement to the chapter entitled, "Resolving Agents and Resolutions in Organic Chemistry" in Volume 6 of "Topics in Stereochemistry," Wiley-Interscience, New York, 1971. In addition to information on resolving agents and methods, critically evaluated optical rotation data on over a thousand compounds is included.

Wyn-Jones, E., "Quantitative Studies of Conformational Energies in Some Cyclic Compounds," *J. Chem. Educ.*, *48*, 402 (1971).

A brief discussion of equilibration, ultrasonic relaxation and nmr methods for determining conformational energies.

HISTORY

Snelders, H. A. M., "The Reception of J. H. van't Hoff's Theory of the Asymmetric Carbon Atom," *J. Chem. Educ.*, *51*, 3 (1974).

A historical account of the difficulties encountered by van't Hoff in gaining acceptance of his theory.

Weyer, J. "A Hundred Years of Stereochemistry - The Principle Development Phases in Retrospect," *Angew. Chemie. Int. Ed.*, *13*, 591 (1974).

An interesting, eight page article on the history of stereochemistry starting with van't Hoff.

INORGANIC CHEMISTRY

Aylett, B. J., "The Stereochemistry of Main Group IV Elements," *Progress in Stereochemistry*, *4*, 213 (1969).

Discussion of the stereochemistry of carbon, silicon, germanium, tin and lead. Emphasis on silicon and tin.

Busch, D. H., "The Stereochemistry of Complex Inorganic Compounds," *J. Chem. Educ.*, *41*, 77 (1964).

A general review of the stereochemistry of coordination complexes.

Fergusson, J. E., *Stereochemistry and Bonding in Inorganic Chemistry*, Prentice-Hall, Inc., Englewood Cliffs, N. J., 1974.

Survey of current bonding theories as they apply to shape of inorganic compounds and a systematic survey of the stereochemistry of inorganic compounds. Assumes some understanding of quantum mechanics and is weak on references.

Hawkins, C. J., *Absolute Configuration of Metal Complexes*, Wiley-Interscience, New York, 1971.

Complete coverage of the subject from nomenclature to spectroscopy. The author uses "absolute configuration" in its broadest sense to cover geometrical isomerism, optical isomerism, and diastereoisomerism.

White, R. F. M., "NMR Spectroscopy and Inorganic Stereochemistry," *Progress in Stereochemistry*, *4*, 167 (1969).

Applications of nmr spectra in inorganic chemistry are discussed for Groups I through VII of the periodic table.

MODELS AND DRAWINGS

Curtin, D. Y., "Stereo Pair Drawings of Crystal Structures Prepared by a Desk Calculator-Computer," *J. Chem. Educ.*, *50*, 775 (1973).

Description of a method for preparing stereo pair drawings, including the mathematical basis, design and operation of the computer program.

Fieser, L. F., "Plastic Dreiding Models," *J. Chem. Educ.*, *40*, 457 (1963).

The advantages of the plastic-aluminum Dreiding models developed by the author are discussed.

Nelson, G. V., "The Preparation and Projection of Inexpensive 35mm Three-Dimensional Slides," *J. Chem. Educ.*, *51*, 47 (1974).

A very brief discussion.

Stong, C. L., "The Amateur Scientist: Molecular Models and an Interferometer that can be Constructed at Modest Cost," *Scientific American*, *228:2*, 110 (February 1973).

Directions for constructing molecular models with inexpensive materials.

Walton, A., "The Use of Models in Stereochemistry," *Progress in Stereochemistry, 4,* 335 (1969).

Space-filling, ball and spoke, skeletal, and home-made models are described, along with suppliers and prices.

NOMENCLATURE

Blackmore, P. F., Williams, J. F., and Clark, M. G., "Biological Asymmetry of Glycerol," *J. Chem. Educ., 50,* 555 (1973).

The recently adopted, correct nomenclature, "Sn-glycero-3-phosphoric acid" is promoted.

Blackwood, J. E., *et al.,* "Unambiguous Specification of Stereoisomerism about a Double Bond," *J. Amer. Chem. Soc., 90,* 509 (1968).

The E-Z system is introduced.

Cahn, R. S., Ingold, C., and Prelog, V., "Specification of Molecular Chirality," *Angew. Chemie, Inter. Ed., 5,* 385 (1966).

Hallmark effort to give every three-dimensional dissymmetric structure a chiral symbol.

Cahn, R. S., "An Introduction to the Sequence Rule," *J. Chem. Educ., 41,* 116 (1964).

The R-S system is presented.

Eliel, E. L., "Recent Advances in Stereochemical Nomenclature," *J. Chem. Educ., 48,* 163 (1971).

A lucid summary of the most significant changes and additions in stereochemical nomenclature since publication of the author's stereochemistry text in 1962. The topics discussed include changes in the R-S system, the E-Z system for olefins, the Beilstein r-system of naming diastereomers in poly-substituted cyclanes, and the enantiotopic-diastereotopic system.

Hanson, K. R., "Applications of the Sequence Rule," *J. Amer. Chem. Soc., 88,* 2731 (1966).

The *pro-R/pro-S* and *re/si* systems are presented.

International Union of Pure and Applied Chemistry, Commission on the Nomenclature of Organic Chemistry, "IUPAC Tentative Rules for the Nomenclature of Organic Chemistry. Section E. Fundamental Stereochemistry," *J. Org. Chem.*, *35*, 2849 (1970).

The definitive publication on stereochemical nomenclature. Terms are prescribed to cover the basic concepts of stereochemistry and the ways in which these terms may be incorporated into the names of individual compounds are defined.

Slocum, D. W., Sugarman, D., and Tucker, S. P., "The Two Faces of D and L Nomenclature," *J. Chem. Educ.*, *48*, 597 (1971).

A critical review of the Fischer D-L nomenclature, including a summary of its historical development and an analysis of the conflicts and ambiguities in the D-L system.

ODOR, TASTE, AND VISION

Amoore, J. E., "Stereochemical and Vibrational Theories of Odor," *Nature*, *233*, 270 (1971). See also *Nature*, *233*, 231 (1971).

A report on quantitative correlations of odor and molecular shape by one of the leading theorists in the field.

Amoore, J. E., Johnston, J. W.,Jr., and Rubin, M., "The Stereochemical Theory of Odor," *Scientific American*, *210:2*, 42 (February 1964).

Evidence that the sense of smell is based on the geometry of molecules is reviewed in this well-illustrated article.

Davies, J. T., "What Makes a Molecule Odorous?", *Nature*, *251*, 97 (1974).

Some recent views, including references, on some of the recently published theories of odor.

Friedman, L., and Miller, J. G., "Odor Incongruity and Chirality," *Science*, *172*, 1044 (1971).

A report that a difference in odor of the enantiomers of carvone is unambiguously demonstrated by chemical interconversion. The authors also note that many enantiomers do not possess different odors.

Guild, W., Jr., "Theory of Sweet Taste," *J. Chem. Educ.*, *49*, 171 (1972).

A short discussion of the molecular features which appear to be common to compounds which taste sweet.

Hubbard, R., and Kropf, A., "Molecular Isomers in Vision," *Scientific American, 216:6,* 64 (June 1967).

A well-illustrated presentation of the roles of the *cis-* and *trans-* forms of retinal, the fundamental molecule of vision.

Murov, S. L., and Pickering, M., "The Odor of Optical Isomers," *J. Chem. Educ., 50,* 74 (1973).

A laboratory experiment in the separation and characterization of *l*-carvone from spearmint oil and *d*-carvone from caraway seed oil is described.

Russell, G. F., and Hills, J. I., "Odor Differences Between Enantiomeric Isomers," *Science, 172,* 1043 (1971).

A report on the odor differences between enantiomers for carvone and related compounds.

OPTICAL ACTIVITY AND PURITY

Bonner, W. A., Kavasmaneck, P. R., Martin, F. S., and Flores, J. J., "Asymmetric Adsorption of Alanine by Quartz," *Science, 186,* 143 (1974).

This probably is the first valid demonstration that optically active quartz can bind one enantiomer of a substance more than the other. The reported extent of asymmetrical preferential adsorption was 1.0 to 1.8 per cent.

Elias, W. E., "The Natural Origin of Optically Active Compounds," *J. Chem. Educ., 49,* 448 (1972).

A critical review of the theories advanced to explain how optical activity in naturally occurring compounds originated.

McCreary, M. D., Lewis, D. W., Wernick, D. L., and Whitesides, G. M., "The Determination of Enantiomeric Purity Using Chiral Zanthanide Shift Reagents," *J. Amer. Chem. Soc., 96,* 1038 (1974).

The direct determination of enantiomeric purity is a problem which is under current investigation. This is one of the more recent articles in this area.

Mowery, D. F., Jr., "Criteria for Optical Activity in Organic Molecules," *J. Chem. Educ. 46,* 269 (1969).

An effort to clarify the reason for optical activity and criteria for prediction of the possible existence of optical isomers.

Newnham, R. E., and Cross, L. E., "Ambidextrous Crystals," *Endeavor, 33,* 18 (1974).

A report of ambidextrous crystals (for example, lead germanate) which have a handedness that can be switched by application of an external electric field.

Pincock, R. E., and Wilson, K. R., "Spontaneous Generation of Optical Activity," *J. Chem. Educ., 50,* 455 (1973).

A critical evaluation of the theory that dissymmetric material or forces must be present in order to generate other dissymmetric material.

Raban, M., and Mislow, K., "Modern Methods for the Determination of Optical Purity," *Topics in Stereochem., 2,* 199 (1967).

Isotopic dilution, kinetic, chromatographic, nmr, and calorimetric methods are discussed.

REACTIONS

Apsimon, J. W., "Aspects of Stereoselective Synthesis," Chapter XVIII in Weissberger, A., ed., *Techniques of Chemistry,* Vol. IV, 2nd ed., ed. K. W. Bentley and G. W. Kirby, Wiley-Interscience, New York, Part III, 1972, p. 251.

Authoritative review of stereoselective processes and discussion of prospects for further advances, 404 references.

Capon, Brian, "Neighboring Group Participation," *Quart. Rev., 18,* 45 (1964).

A comprehensive review.

Goller, Edwin J., "Stereochemistry of Carbonyl Addition Reactions," *J. Chem. Educ., 51,* 182 (1974).

Attempts to explain the steric course of nucleophilic addition reactions in simple carbonyl systems are reviewed. The unifying concepts that have been proposed are discussed.

Morrison, J. D. and Mosher, H. S., *Asymmetric Organic Reactions,* Prentice-Hall, Inc., Englewood Cliffs, New Jersey, 1971.

A complete critical overview of asymmetric synthesis, with extensive tabulation and correlation of examples.

Raber, D. J., and Harris, J. M., "Nucleophilic Substitution Reactions at Secondary Carbon Atoms," *J. Chem. Educ., 49,* 60 (1972).

Important recent advances in mechanistic studies of these reactions are presented with emphasis on criteria for the assignment of mechanism.

Scott, J. W., and Valentine, D., Jr., "Asymmetric Synthesis," *Science, 184,* 943 (1974).

This review article is concerned primarily with work reported since 1969, that is, since the book by Morrison and Mosher which is cited in this section.

SPECTROSCOPY AND SYMMETRY

Beychok, S., "Circular Dichroism of Biological Macromolecules," *Science, 154,* 1288 (1966).

A lucid discussion of the nature of ORD and CD spectra and of their use in studying the conformations of proteins and nucleic acids in solution.

Carlos, J. L., Jr., "Molecular Symmetry and Optical Inactivity," *J. Chem. Educ., 45,* 248, (1968).

The S_n criterion for optical inactivity is discussed along with some apparent exceptions to this rule.

Crabbé, P., *ORD and CD in Chemistry and Biochemistry; An Introduction,* Academic Press, New York, 1972.

An extensive presentation of the origin of ORD and CD effects and their applications in solving problems in chemistry and biochemistry.

Donaldson, J. D., and Ross, S. D., *Symmetry and Stereochemistry,* Wiley-Interscience, New York, 1972.

Symmetry elements and operations, point group symmetry, space group symmetry, and group theory are discussed.

Kennewell, P. D., "Applications of the Nuclear Overhauser Effect in Organic Chemistry," *J. Chem. Educ., 47,* 278 (1970).

A brief presentation of the use of this effect for precise assignments of nmr signals to particular protons and elucidation of the structure of complex molecules.

Mislow, K., and Raban, M., "Stereoisomeric Relationships of Groups in Molecules," *Topics in Stereochem., 1,* 1 (1967).

Summary of symmetry methods, equivalent-enantiotopic-diastereotopic systems, and nmr shifts and coupling constants.

Orchin, M., and Jaffe, H. H., *Symmetry, Orbitals and Spectra,* Wiley-Interscience, New York, 1971.

An advanced text which relates symmetry, molecular orbital theory and spectroscopy.

Peterson, M. R., and Wahl, G. H., "Lanthanide NMR Shift Reagents, A Powerful New Stereochemical Tool," *J. Chem. Educ.*, *49*, 790 (1972).

A brief readable discussion of the use of paramagnetic reagents to separate resonances that would otherwise overlap.

Sanders, J. K. M., and Williams, D. H., "Shift Reagents in NMR Spectroscopy," *Nature*, *240*, 385 (1972).

An excellent review covering the mechanism of the shifts, the nature of the interactions, applications, and implications for biology.

Slayter, E. M., *Optical Methods in Biology*, Wiley-Interscience, New York, 1970.

An excellent source for those interested in learning more about the chiral properties of light.

Wilson, N. K., and Stothers, J. B., "Stereochemical Aspects of Carbon-13 NMR Spectroscopy," *Topics in Stereochem.*, *8*, 1 (1974).

A comprehensive review of the use of carbon-13 nmr spectra for elucidating the structure of carbon-containing compounds.

Index

HETERICK MEMORIAL LIBRARY
541.223077 G455s
onuu
Giese, Roger W./Stereochemistry : an int
3 5111 00054 6782